184
Advances in Anatomy Embryology and Cell Biology

K. G. Usunoff · A. Popratiloff ·
O. Schmitt · A. Wree

Functional Neuroanatomy of Pain

With 19 Figures

Springer

Kamen G. Usunoff, MD

Department of Anatomy and Histology
Medical University – Sofia
2. Sv. G. Sofiiski ST.
1431 Sofia
Bulgaria

e-mail: uzunoff@medfac.acad.bg

Anastas Popratiloff, MD

Department of Anatomy and Cell Biology
George Washington University Medical Center
Washington, DC 20037
USA

Oliver Schmitt, MD
Andreas Wree, MD

Institut für Anatomie
Universität Rostock
P.O. Box 100888
18055 Rostock
Germany

e-mail: andreas.wree@med.uni-rostock.de

ISSN 0301-5556
ISBN-10 3-540-28162-2 Springer Berlin Heidelberg New York
ISBN-13 978-3-540-28162-7 Springer Berlin Heidelberg New York

Springer is a part of Springer Science+Business Media
springeronline.com

Printed in Germany

Editor: Simon Rallison, Heidelberg
Desk editor: Anne Clauss, Heidelberg
Production editor: Nadja Kroke, Leipzig
Cover design: design & production GmbH, Heidelberg
Typesetting: LE-TEX Jelonek, Schmidt & Vöckler GbR, Leipzig
Printed on acid-free paper SPIN 11533467 27/3150/YL – 5 4 3 2 1 0

List of Contents

Abbreviations

(The abbreviations apply to all figures.)

III	Third ventricle
AA	Axo-axonal terminal
ACC	Anterior cingulate cortex
AMPA	α-Amino-3-hydroxy-5-methyl-4-isoxazolepropionic acid
AP	Area postrema
BDNF	Brain-derived neurotrophic factor
Bi	Midline nucleus of Bischoff
CCI	Chronic constriction injury
CCK	Cholecystokinin
CGRP	Calcitonin gene-related peptide
CL	Nucleus centralis lateralis
Cu	Cuneate nucleus
C1, C2	Central terminals of type 1 or type 2 glomerulus
D	Dendrite
DCN	Dorsal column nuclei
DH	Dorsal horn
DT	Dome-shaped terminal
EPSP	Excitatory postsynaptic potential
EM	Electron microscopy
FB	Fast Blue
FGF-2	Fibroblast growth factor-2
fMRI	Functional magnetic resonance imaging
FRAP	Flour-resistant acid phosphatase
GABA	γ-Aminobutyric acid
GDNF	Glial cell line-derived neurotrophic factor
GluR1	AMPA receptor subunits GluR1
GluR2	AMPA receptor subunits GluR2
Gr	Gracile nucleus
HZ	Herpes zoster
IC	Insular cortex
ION	Infraorbital nerve
LCN	Lateral cervical nucleus
LM	Light microscopy

LSN	Lateral spinal nucleus
MDH	Medullary dorsal horn
MD	Mediodorsal thalamic nuclei
MDvc	Medial thalamus, ventrocaudal part
NGF	Nerve growth factor
NMDA	*N*-methyl-D-aspartate
NMDAR1	NMDA receptor subunit 1
NMDAR2	NMDA receptor subunit 2
NKA	Neurokinin A
NK1	Neurokinin 1
NO	Nitric oxide
NOS	Nitric oxide synthase
NP	Neuropathic pain
NPY	Neuropeptide Y
PA	Primary afferent (neuron)
PC	Prefrontal cortex
RF	Reticular formation
PAG	Periaquaductal gray
PET	Positron emission tomography
PHN	Postherpetic neuralgia
Po	Posterior nuclear complex
Pom	Posterior nuclear complex, medial part
PTN	Principal trigeminal nucleus
SC	Spinal cord
SG	Spinal (dorsal root) ganglia
SHT	Spinohypothalamic tract
SMT	Spinomesencephalic tract
Sol	Nucleus solitarius
SP	Substance P
SPbT	Spinoparabrachial tract
SRT	Spinoreticular tract
STN	Spinal trigeminal nucleus
STNc	Spinal trigeminal nucleus, caudal part (subnucleus caudalis)
STNi	Spinal trigeminal nucleus, interpolar part (subnucleus interpolaris)
STNo	Spinal trigeminal nucleus, oral part (subnucleus oralis)
STrT	Spinal trigeminal tract
STT	Spinothalamic tract
SI	Primary somatosensory cortex
SII	Secondary somatosensory cortex
TG	Trigeminal ganglion
THT	Trigeminohypothalamic tract
TTT	Trigeminothalamic tract
VIP	Vasoactive intestinal polypeptide
VL	Nucleus ventralis lateralis

VMpo	Nucleus ventralis medialis, posterior part
VPI	Nucleus ventralis posterior inferior
VPL	Nucleus ventralis posterior lateralis
VPLc	Nucleus ventralis posterior lateralis, caudal part
VPLo	Nucleus ventralis posterior lateralis, oral part
VPM	Ventral posteromedial thalamic nucleus
VR1, VRL1	Vanilloid receptors 1 and L1
VZV	Varicella-zoster virus

1 Introduction

Pain is defined by the International Association for the Study of Pain (IASP) as an unpleasant sensory and emotional experience associated with actual or potential tissue damage, or described in terms of such damage or both. Pain is an unpleasant but very important biological signal for danger. Nociception is necessary for survival and maintaining the integrity of the organism in a potentially hostile environment (Hunt and Mantyh 2001; Scholz and Woolf 2002). Pain is not a monolithic entity. It is both a sensory experience and a perceptual metaphor for damage (i.e., mechanically, by infection), and it is activated by noxious stimuli that act on a complex pain sensory apparatus.

However, sustained or chronic pain can result in secondary symptoms (anxiety, depression), and in a marked decrease of the quality of life. This spontaneous and exaggerated pain no longer has a protective role, but pain becomes a ruining disease itself (Basbaum 1999; Dworkin and Johnson 1999; Woolf and Mannion 1999; Dworkin et al. 2000; Hunt and Mantyh 2001; Scholz and Woolf 2002). If pain becomes the pathology, typically via damage and dysfunction of the peripheral and central nervous system, it is termed "neuropathic pain."

Here, we present an updated review of the functional anatomy of normal and neuropathic pain.

2 Functional Neuroanatomy of the Pain System

2.1 Primary Afferent Neuron

The primary afferent (PA) neuron is the pseudounipolar cell, localized in spinal (dorsal root) ganglia (SG), and in the sensory ganglia of the 5th, 7th, 9th, and 10th nerves (for reviews see Scharf 1958; Duce and Keen 1977; Brodal 1981; Willis 1985; Zenker and Neuhuber 1990; Willis and Coggeshall 1991; Hunt et al. 1992; Lawson 1992; Waite and Tracey 1995; Usunoff et al. 1997; Waite and Aschwell 2004). The perikarya of the PA neurons are round, oval, or elliptical. The neurons lack dendritic processes and generally lack direct synaptic input to the soma (Feirabend and Marani 2003). The Nissl substance is abundant but finely dispersed. In old individuals, large accumulations of lipofuscin are regularly observed. Feirabend and Marani (2003) summarized the functional aspects of the dorsal root ganglia: "It appears that the DRG cell bodies are electrically excitable, lack a blood brain barrier and some are able to fire repetitively. The first feature may be important for both propagation of impulses along the T junction and feed back regulation of sensory endings. The second aspect suggests a role as chemical sensor and the third property may be responsible for generating background sensation of the awareness of the body scheme." The cell body emits a single process (crus commune) that bifurcates in a peripheral and central process. Frequently, and especially in the larger neurons, the crus commune is highly coiled (Ramon y Cajal 1909); this is referred to as the glomerular segment. The central process, usually thinner than the peripheral one (Rexed and Sourander 1949), enters the CNS, and the peripheral process (morphologically an axon, functionally a dendrite) runs in the peripheral nerve to its sensory innervation zone. The peripheral specialized transductive ending serves as part of a sense organ complex or as the sense organ itself as is the case with the free nerve ending.

The diameter of the pseudounipolar perikarya varies from 15 to 110 μm. Two basic types are generally recognized: large, light A cells and small, dark B cells. The cytoplasm of the large cells is rather pale and unevenly stained due to aggregations of Nissl substance interspersed with light staining regions that contain microtubules and a large amount of neurofilaments. The small cells appear dark mainly because of the densely packed cisternae of granular endoplasmic reticulum and few neurofilaments. The largest A cells are the typical proprioceptor neurons, and the small B cells are the typical nociceptor neurons (Harper and Lawson 1985; Sommer et al. 1985; LaMotte et al. 1991; Willis and Coggeshall 1991; Truong et al. 2004). The neurons in the trigeminal ganglion (TG) are similarly distinguished in light and dark cells (Capra and Dessem 1992; Waite and Tracey 1995; Usunoff et al. 1997; Waite and Ashwell 2004). Attempts have been made to classify the two populations of PA neurons further into physiological, anatomical, ultrastructural, and immunocytochemical terms (Sommer et al. 1985; Lawson et al. 1987, Lawson 1992, 2002; Schoenen and Grant 2004). Some studies suggest that a single PA neuron may give rise to more than one peripheral branch, and more than one centrally projecting branch (Langford and Coggeshal 1981; Chung and Coggeshal 1984; Alles and Dom 1985; Laurberg and Sorensen 1985; Coggeshall 1986; Nagy et al. 1995; Russo and Conte 1996; Sameda et al. 2003). This question is of interest from a clinical point of view because the possible branching of peripheral processes has bearing on the problem of referred pain (Coggeshall 1986; Schoenen and Grant 2004).

There are numerous studies on the number and size of PA neurons of the SG in various species revealing not only large species differences but also significant interindividual variations (Avendano and Lagares 1996; Mille-Hamard et al. 1999; Farel 2002; Tandrup 2004). Ball et al. (1982) examined the TG from 64 human subjects from 2 months to 81 years old; the mean neuronal count was 80,600 with no significant age or sex difference. However, they reported striking variation in individual samples (range 20,000–157,000). According to a recent investigation, the human TG comprises approximately 20,000–35,000 neurons (La Guardia et al. 2000).

The neurotransmitter of the PA cells is the amino acid glutamate, the most typical fast-acting central excitatory transmitter (Weinberg et al. 1987; De Biasi and Rustioni 1988; Rustioni and Weinberg 1989; Clements et al. 1991; Westlund et al. 1992; Broman et al. 1993; Broman 1994; Valtschanoff et al. 1994; Salt and Herrling 1995; Keast and Stephensen 2000; Meldrum 2000; Lazarov 2002; Hwang et al. 2004; Tao et al. 2004). The glutamate acts postsynaptically on three families of ionotropic receptors, named after their preferred agonists, *N*-methyl-D-aspartate (NMDA), α-amino-3-hydroxy-5-methyl-4-isoxazolepropionic acid (AMPA), and kainate. These receptors all incorporate ion channels that are permeable to cations, although the relative permeability to Na^+ and Ca^{++} varies according to the family and the subunit composition of the receptor (Hollmann et al. 1989; Yoshimura and Jessel 1990; Furuyama et al. 1993; Tölle et al. 1993, 1995; Hollmann and Heinemann 1994; Petralia et al. 1994, 1997; Tachibana et al. 1994; Popratiloff et al. 1996a, b; Ruscheweyh and Sandkühler 2002; Szekely et al. 2002). More recently, also glutamate metabotropic receptors were discovered. They are G-proteins linked and operate by releasing second messengers in the cytoplasm, or by influencing ion channels through release of G-protein subunits within the membrane (Schoepp and Conn 1993; Pin and Duvoisin 1995; Conn and Pin 1997). Glutamate is released from the peripheral terminals of PA nociceptors in the skin and joints during sensory transduction presumably as an initiating event in neurogenic inflammation (Lawand et al. 1997; Carlton and Coggeshall 1999; Carlton et al. 2001; Willis and Westlund 2004).

Especially the B cells contain, besides glutamate, various neuropeptides: substance P (SP), calcitonin gene-related peptide (CGRP), galanin, neuropeptide Y (NPY), neurokinin A (NKA), somatostatin, cholecystokinin (CCK), bombesin, vasoactive intestinal polypeptide (VIP), dynorphin, enkephalin, etc. (Rustioni and Weinberg 1989; Willis and Coggeshall 1991; Lawson 1992; Levine et al. 1993; Broman 1994; Ribeiro-da-Silva 1995; Wiesenfeld-Hallin and Xu 1998; Edvinsson et al. 1998; Todd 2002; Waite and Ashwell 2004; Willis and Westlund 2004). Two or more peptides may be colocalized in the same PA. The proportions of peptidergic SG cells that contain a particular peptide may differ depending on the type of peripheral nerve. CGRP is found in 50% of skin afferents, in 70% of muscle afferents, and in practically all visceral afferents. SP is found in 25% of skin afferents, in 50% of muscle afferents, and in more than 80% of visceral afferents. However, somatostatin is lacking in visceral afferents but is present in a small number of somatic afferents (Willis and Westlund 2004). According to Ambalavanar et al. (2003) from the cutaneous PA neurons in the rat's TG, 26% contain CGRP, 5% SP, and 1% somatostatin. In the SG, the quantity of SP-containing neurons (10%–29% of the cutaneous afferent population) is considerably higher (O'Brien et al. 1989; Hökfelt 1991; Willis and Coggeshall 1991; Perry and Lawson 1998; see also Lazarov 2002). Most cells containing SP seem to be nociceptive neurons with high thresholds (Lawson et al. 1997). In the SG (Yang et al. 1998), the percentage of CGRP-immunoreactive neurons is smaller in females that in males. In guinea pigs, the CGRP expression is detected in under half the nociceptive neurons, and is not limited to nociceptive neurons (Lawson et al. 2002). It seems likely that the peptides are neuromodulators that act in concert with the fast-acting neurotransmitter glutamate, either enhancing or diminishing its action (Levine et al. 1993; Willis et al. 1995; Besson 1999; McHugh and McHugh 2000).

The brain-derived neurotrophic factor (BDNF) meets many of the criteria to establish it as a neurotransmitter/neuromodulator in small diameter nociceptive

PA neurons, localized in dense core synaptic vesicles (McMahon and Bennett 1999; Mannion et al. 1999; Pezet et al. 2002) and is released by the PAs terminating in the superficial laminae of the dorsal horn (DH).

The gaseous transmitter nitric oxide (NO) is synthesized by the enzyme nitric oxide synthase (NOS) in some PA cells of the SG, and in the sensory ganglia of the cranial nerves (Morris et al. 1992; Aoki et al. 1993; Terenghi et al. 1993; Alm et al. 1995; Dun et al. 1995; Lazarov 2002; Thippeswamy and Morris 2001, 2002; Luo et al. 2004). NO is found mainly in the small sensory neurons (Zhang et al. 1993b; Vizzard et al. 1994; Lazarov and Dandov 1998; Rybarova et al. 2000) and coexists with CGRP, sometimes also with SP and galanin (Zhang X et al. 1993a; Majewski et al. 1995; Edvinsson et al. 1998; Rybarova et al. 2000). In the human TG, the coexistence of NO and CGRP is less pronounced (Tajti et al. 1999).

The peripheral processes of the nociceptive PA cells terminate generally as thin fibers of two types: Aδ (Group III), and C (Group IV) (Perl 1996; Bevan 1999; Basbaum and Jessel 2000; Lewin and Moshourab 2004; Willis and Westlund 2004). The Aδ-fibers are thinly myelinated, with a diameter of 1–3 μm and a conduction velocity of 5–30 m/s. More rapidly conducting nociceptive A-fibers (up to 51 m/s) have been described (Treede et al. 1995). The C-fibers are unmyelinated, with a diameter of approximately 1 μm and with a conduction velocity of 0.5–2 m/s. Goldschneider (1881) was the first to propose the existence of two pains, later universally recognized (Hassler 1960; Bowsher 1978; Craig 2003a, d). The first pain (pinprick sensation) is typical for threat of tissue damage. It is rapidly conducted to consciousness and well localized. The second pain occurs when tissue damage has already taken place. It is slowly conducted and poorly localized (Basbaum and Jessel 2000; Julius and Basbaum 2001).

Nociceptors respond maximally to overtly damaging stimuli, although they generally also respond, but less vigorously, to stimuli that are merely threatening (Willis and Westlund 2004). Stimulation of cutaneous Aδ-nociceptors leads to pricking pain, whilst stimulation of C-nociceptors leads to burning or dull pain (Campbell and Meyer 1996; Perl 1996; Willis and Westlund 1997, 2004; Millan 1999; Raja et al. 1999). The peripheral processes of nociceptive PA neurons terminate as free nerve endings (Cauna 1980; Kruger et al. 1981, 2003a, b; Halata and Munger 1986; Kruger 1988, 1996; Munger and Ide 1988; Heppelmann et al. 1995; Messlinger 1996; Petruska et al. 1997; Fricke et al. 2001). The nociceptor terminal differs from other sense organs in responding more vigorously to successive identical stimuli, a process called sensitization. This contrasts with the reduced responsiveness to successive stimuli known as adaptation—displayed by all other sensory transduction systems (Kruger et al. 2003b). Nociceptors, in contrast to modality specificity of other sense organs, are apparently responsive to mechanical, chemical and thermal perturbations, accounting for their common designation as polymodal (Kruger 1996).

The sensory endings of group III (Aδ) and group IV (C) are characterized by varicose segments, the sensory beads, described by Ramon y Cajal (1909) in the cornea. They measure 5–12 μm in length in group III and 3–8 μm in group IV fibers (Messlinger 1996). The free nerve endings contain clusters of small clear vesicles, dense core vesicles, membranous strands of smooth endoplasmic reticulum, mitochondria, and sometimes glycogen granules (Messlinger 1996; Kruger et al. 2003a, b). The nociceptors, except the free endings, are incompletely surrounded by modified Schwann cells. In particular, their beads exhibit free areas where the axolemma is separated from the surrounding tissue by the basal lamina only. The axoplasm that underlies the bare areas of axolemma shows a faint filamentous substructure and appears more electron-dense (Messlinger 1996). A high concentration of axonal mitochondria may be correlated with energy consumption and hence the activity of the sensory endings (Heppelmann et al. 1994). Probably, the sensory beads represent the receptive sites of the sensory endings (Andres and von Düring 1973; Chouchkov 1978; Munger and Halata 1983; Messlinger 1996).

The free nerve endings contain SP, CGRP, and NKA (Gibbins et al. 1987; Dalsgaard et al. 1989; Micevych and Kruger 1992; Dray 1995; Kruger 1996; Holland et al. 1998), and the sensory endings in the cornea contain also galanin (Marfurt et al. 2001; Müller et al. 2003). However, the neuropeptides, released by the endings, do not have a neurotransmitter function (for a discussion on the noceffector concept, see Kruger 1996).

2.2
Distribution of Nociceptor Peripheral Endings

The free nerve endings are to be found throughout the body, mainly in the adventitia of small blood vessels, in outer and inner epithelia, in connective tissue capsules, and in the periosteum. They are most densely arranged in the cornea, dental pulp, skin and mucosa of the head, skin of the fingers, parietal pleura, and peritoneum.

The two main types of nociceptors in the skin are Aδ mechanical and C polymodal nociceptors (Willis and Westlund 2004), although other types of nociceptors have also been described (Davis et al. 1993). Within the dermis, the afferent fiber gives off several branches that penetrate the basal lamina and extend into the epidermis. As a rule, the myelin sheath ends within the dermis. Most large axons lose their myelin sheaths and perineurium before reaching the papillary layer of the dermis, with the exception of the axons innervating Merkel cells, although those also become unmyelinated before penetrating the epidermis (Iggo and Muir 1969; Kruger et al. 1981; Halata et al. 2003). Cauna (1973) described an elaborate cluster of unmyelinated fibers entering the papillary layer of human hairy skin as a free "penicillate ending". Terminals that penetrate the epidermis for a considerable distance (to the stratum granulosum) have been reported in studies, utilizing methylene blue or silver stainings (Woolard 1935). In the beginnings of ultrastructural examination, numerous reports on the electron microscopic image of the skin receptors appeared (Halata 1975; Andres and von Düring 1973; Cauna 1973, 1980; Chouchkov 1978; Kruger et al. 1981). Even in recent papers (Kruger 1996; Kruger and Halata 1996; Messlinger 1996; Kruger et al. 2003a, b) the authors are careful in the description of the intraepithelial run of the free nerve endings. As the axon-Schwann cell complex approaches the basal epidermis, the thin Schwann cell basal lamina merges with the thicker epidermal basal lamina. The axon penetrating the epidermis is accompanied by thin Schwann cell processes which follow its course until a single axonal profile is completely enveloped by keratinocytes, without junctional specializations (Kruger et al. 1981, 2003b).

The Meissner corpuscles are widely regarded as low-threshold mechanoreceptors. However, Pare et al. (2001) showed that Meissner corpuscles are multiafferented receptor organs that may have also nociceptive capabilities. In the Meissner corpuscles of glabrous skin of monkey digits they found that the Aα-β-fibers are closely intertwined with endings of peptidergic C-fibers (SP and CGRP). These intertwined endings are segregated into zones containing nonpeptidergic C-fibers expressing immunoreactivity for vanilloid receptor 1.

The enormous number of free nerve endings in the cornea and the lack of any encapsulated receptors were demonstrated by Ramon y Cajal as early as 1909. The innervation density is 300–600 times that of the skin (Rozsa and Beuerman 1982). The number of PA neurons in the TG, that send their peripheral processes in the ophthalmic nerve is modest (La Vail et al. 1993); however, a single corneal sensory neuron in the rabbit support approximately 3,000 individual nerve endings (Marfurt et al. 1989; Belmonte and Gallar 1996; Müller et al. 2003). Both myelinated Aδ and unmyelinated C-fibers are present in the peripheral cornea but the central cornea is innervated by unmyelinated fibers. The latter penetrate Bowman's membrane and terminate between the epithelial cells (Müller et al. 2003; Waite and Ashwell 2004; Guthoff et al. 2005).

Human premolars receive about 2,300 axons at the root apex, and 87% of these fibers are unmyelinated. Most apical myelinated axons are fast conducting Aδ-fibers with their receptive fields located at the pulpal periphery and inner dentin. These fibers are probably activated by a hydrodynamic mechanism and conduct impulses that are perceived as a short, well-localized sharp pain. Most C-fibers are slow-conducting fine afferents with their receptive fields located in the pulp and transmit impulses that are experienced as dull, poorly localized and lingering pain (Nair 1995; Waite and Ashwell 2004). Free nerve endings in mature teeth are found in the peripheral plexus of Rashkow, the odontoblastic layer, the predentin, and the dentin. The endings are most numerous in the regions near the tip of the pulp horn, where more than 40% of the dentinal tubules can be innervated (Byers 1984). Endings can extend for up to 200 μm into the dentinal tubules in both monkey and human teeth, particularly near the cusps of the crown (Byers and Dong 1983; Waite and Ashwell 2004). The periodontal ligament is rich in free nerve endings. The periodontal pain is usually well localized and exacerbated by pressure (Waite and Ashwell 2004).

In the muscles, the free nerve endings are found in the adventitia of the blood vessels, but also between muscle fibers, in the connective tissue of the muscle and in the tendons (Andres et al. 1985). The small myelinated afferent fibers in the muscles have conduction velocities from 2.5–20 m/s, and the unmyelinated fibers less than

2.5 m/s. Of all of the small myelinated and unmyelinated fibers, approximately 40% were believed to be nociceptors (Marchettini et al. 1996; Mense 1996). Bone has a rich sensory innervation; the density of nociceptors in the periosteum is high, whereas nerve fibers in the mineralized portion of the bone are less concentrated and are associated with blood vessels in Volkman and Haversian canals (Bjurholm et al. 1988; Hill and Elde 1991; Hukkanen et al. 1992; Mach et al. 2002). Nociceptors in the joint are located in the capsule, ligaments, bone, articular fat pads, and perivascular sites, but not in the joint cartilage (Heppelmann et al. 1990; Hukkanen et al. 1992; Halata et al. 1999). The free nerve endings in the cruciate ligaments are found subsynovially, and are seen also between collagen fibers, close to blood vessels. However, at least part of the latter fibers appear to be efferent sympathetic fibers and not nociceptors (Halata et al. 1999). The branched, terminal tree of the unmyelinated fibers has a "string of beads" appearance which probably represent multiple receptive sites in the nerve ending (Heppelmann et al. 1990; Schmidt 1996).

In the healthy back, only the outer third of the annulus fibrosus of the intervertebral disk is innervated (see Coppes et al. 1990, 1997; Freemont et al. 1997). Lower back pain was studied in diseased lumbar intervertebral discs and was for the first time reported to be related to ingrowth of nociceptive fibers by Coppes et al. (1990, 1997). This finding was confirmed in 46 samples of diseased intervertebral disks (Freemont et al. 1997). Both groups characterized this ingrowth and extension of the neuronal disk network by the nociceptive neurotransmitter substance P. It is now well established that a change of the innervation of the disk is the morphological substrate for discogenic pain.

There are two classes of nociceptors in viscera (Cervero 1994). The first class is composed of "high-threshold" receptors that respond to mechanical stimuli within the noxious range. Such are found within different organs: gastrointestinal tract, lungs, ureters and urinary bladder, and heart (Cervero 1996; Gebhart 1996). The second class has a low threshold to natural stimuli and encodes the stimulus intensity in the magnitude of their discharges: "intensity-encoding" receptors. Both receptor types are concerned mainly with mechanical stimuli (stretch) and are involved in peripheral encoding of noxious stimuli in the organs (Cervero and Jänig 1992). The cardiac receptors are the peripheral processes of the pseudounipolar PA neurons, located in the SG and the ganglion inferius n. vagi. The sympathetic afferents are considered solely responsible for the conduction of pain arising in the heart. However, Meller and Gebhart (1992) suggest that afferent fibers of the vagus nerve might also contribute to the cardiac pain. The vagus nerve is largely responsible for the pain conduction arising in the lung. Klassen et al. (1951) demonstrated that the burning sensation caused by an endobronchial catheter can be abolished by vagal block. In general, solid organs are least sensitive, whereas the serous membranes, covering the viscera are most sensitive to nociceptive stimuli (Giamberardino and Vecchiet 1996).

Except for avascular structures, such as cornea, skin, and mucosa epithelia, nociceptors are adjacent to capillaries and mast cells (Kruger et al. 1985; Dalsgaard et al. 1989; Heppelmann et al. 1995; Messlinger 1996). This triad is a functional nociceptive response unit, which is sensitive to tissue damage (Kruger 1996; McHugh and McHugh 2000). The firing of nociceptors at the site of tissue injury causes release of vesicles containing the peptides SP, NKA, and CGRP, which act in an autocrine and paracrine manner to sensitize the nociceptor and increase its rate of firing (Holzer 1992; Donnerer et al. 1993; Dray 1995; Kruger 1996; Cao et al. 1998; Holzer and Maggi 1998; Millan 1999; McHugh and McHugh 2000). Cellular damage and inflammation increase concentrations of chemical mediators such as histamine, bradykinin, and prostaglandins in the area surrounding functional pain units. These additional mediators act synergistically to augment the transmission of nociceptive impulses along sensory afferent fibers (McHugh and McHugh 2000). In addition to familiar inflammatory mediators, such as prostaglandins and bradykinin, potentially important, pronociceptive roles have been proposed for a variety of "exotic" species, including protons, purinergic transmitters, cytokines, neurotrophins (growth factors), and NO (Mannion et al. 1999; Millan 1999; Boddeke 2001; Willis 2001; Mantyh et al. 2002; Scholz and Woolf 2002). Physiological pain starts in the peripheral terminals of nociceptors with the activation of nociceptive transducer receptor/ion channel complexes inducing changes in receptor potential, which generate depolarizing currents in response to noxious stimuli (Woolf and Salter 2000). In PA neurons, the transducer proteins that respond to extrinsic or intrinsic irritant chemical stimuli are selectively expressed (McCleskey and Gold 1999; and references therein). The noxious heat transducers include the vanilloid receptors VR1 and VRL1 (Caterina et al. 1997, 1999; Tominaga et al. 1998; Guo et al. 1999; Welch et al. 2000; Caterina and Julius 2001; Michael and Priestly 1999; Valtschanoff et al. 2001; Hwang et al. 2003). VR1 are on the terminals of many unmyelinated and some finely myelinated nociceptors and respond to capsaicin, heat, and low pH (Holzer 1991; Caterina et al. 1997, 2000; Helliwell et al. 1998; Tominaga et al. 1998). On the other hand, VRL1 are on PAs with myelinated axons, have a high heat threshold, and do not respond to capsaicin and low pH (Caterina et al. 1999). mRNA for VR1 has been shown to be widely distributed in the brains of both rats and humans (Mezey et al. 2000), so that the role of these receptors in response to painful stimuli may be much more complex than previously thought.

There are nociceptors that under normal circumstances are inactive and rather unresponsive. Such nociceptors were first detected in the knee joint and were called "silent" or "sleeping" by Schaible and Schmidt (1983a, b). Inflammation leads to sensitization of these fibers, they "awaken" and become much more sensitive to peripheral stimulation (Schaible and Schmidt 1985, 1988; Segond von Banchet et al. 2000). Later, "silent" nociceptors were described also in cutaneous and visceral nerves (Davis et al. 1993; McMahon and Koltzenberg 1994; Schmidt et al. 1995, 2000; Snider and McMahon 1998; Petruska et al. 2002).

2.3
Termination in the Spinal Cord and Spinal Trigeminal Nucleus

As central processes of the SG neurons approach the dorsal root entry zone, the fine, nociceptive axons become segregated in lateral portions of the rootlets and enter lateral portions of the DH, passing through fasciculus dorsolateralis Lissaueri (Ranson 1913; Kerr 1975b; Light and Perl 1979a; Brown 1981; Schoenen and Faull 1990; Willis and Coggeshall 1991; Carlstedt et al. 2004). At the junction between spinal cord (SC) and roots, there is a profound redistribution and reorganization of nerve fibers (Fraher 1992, 2000; Carlstedt et al. 2004). The transitional zone is the most proximal free part of the root, which in one and the same cross-section contains both CNS and PNS tissue. The PNS compartment contains astrocytic processes that extend from the CNS compartment forming a fringe among the nerve fibers. The CNS compartment is dominated by numerous astrocytes, while oligodendrocytes and microglia are rare. The myelinated fiber change from PNS to CNS type of organization occurs in a transitional node of Ranvier situated at the proximal end of a glial fringe cul-de-sac at the PNS-CNS borderline.

The nociceptive fibers terminate primarily in the most dorsally located laminae of Rexed (Rexed 1952, 1954, 1964). These comprise lamina I (nucleus postero-marginalis) and lamina II (substantia gelatinosa Rolandi); the Aδ-fibers terminate in laminae I and V, and C-fibers in laminae I and II. The large mechanoreceptive Aβ-axons reach laminae III–VI (Light and Perl 1979a, b; Light et al. 1979; Ralston 1979; Ralston and Ralston 1979; Perl 1996; Willis 1985; Menetrey et al. 1989; Willis and Coggeshall 1991; Hunt et al. 1992; Molander and Grant 1995; Ribeiro-da-Silva 1995; Craig 1996a; Han et al. 1998; Morris et al. 2004). Lamina I is with low neuronal density and contains small, medium-sized, and large neurons. The latter, often called marginal cells of Waldeyer are rich in granular endoplasmic reticulum and other organelles (Ralston 1979). They are usually elongated and the three main perikaryal types are fusiform, pyramidal, and multipolar (Gobel 1978a; Lima and Coimbra 1991; Lima et al. 1991; Zhang ET et al. 1996; Zhang and Craig 1997; Han et al. 1998). Based on responses to natural cutaneous stimuli, there are three major types of lamina I neurons (Craig 1996a): (a) nociceptive-specific neurons that respond only to noxious mechanical or heat stimuli, (b) polymodal nociceptive neurons that respond to noxious heat, pinch, and cold, (c) thermoreceptive-specific neurons that respond to innocuous cooling and are inhibited by warming the skin. The nociceptive-specific neurons are dominated by Aδ-fiber input and can respond tonically to a maintained noxious mechanical stimulus, so they may be important for the “first pain” (Craig 2000). The polymodal nociceptive cells are dominated by C-fiber input and are important for the “second pain.” Han et al. (1998) have shown by means of intracellular labeling that the nociceptive-specific neurons are fusiform, the polymodal nociceptive neurons are multipolar, and the thermoreceptive-specific neurons are pyramidal. Later, Andrew and Craig (2001) identified “itch-specific” lamina I neurons, which are selectively sensitive to histamine. Approximately 80% of lamina I neurons express the NK1 receptor (Todd et al. 2000). Substance P in the PAs acts on the neurokinin 1 (NK1) receptor, which is concentrated in lamina I (Marshall et al. 1996; Todd et al. 1998, 2002; Yu et al. 1999; Cheunsuang and Morris 2000; Mantyh and Hunt 2004; Morris et al. 2004).

Lamina II contains densely packed small cells, with a very low amount of perikaryal cytoplasm but relatively rich dendritic tree (Ralston 1979; Schoenen and Faull 1990, 2003; Ribeiro-da-Silva 1995). Two neuronal types called islet cells and stalked cells are to be distinguished (Gobel 1978b; Todd and Lewis 1986), and in humans, Schoenen and Faull (1990) describe four types: islet, filamentous, curly, and stellate neurons. In lamina II neurons coexist two "classical" inhibitory transmitters: the amino acids γ-aminobutyric acid (GABA) and glycine, and GABA is further co-expressed with the neuropeptides methionine enkephalin and neurotensin (Todd and Sullivan 1990; Todd et al. 1992; Todd and Spike 1993). As originally described by Rexed (1952, 1954) in the cat, lamina II might be subdivided into outer and inner zones. In the outer zone, the neurons are slightly smaller and more tightly packed than in the inner zone. In the rat, Ribeiro-da-Silva (1995) further subdivided lamina II in sublaminae A, Bd, and Bv. In humans, the separation between the outer and the inner zone is much less clear (Schoenen and Faull 1990). It has been postulated that the substantia gelatinosa may function as a controlling system modulating synaptic transmission from PA neurons to secondary sensory systems (Melzack and Wall 1965; Wall 1978; LeBars et al. 1979a, b; Light et al. 1979; Moore et al. 2000). Originally, lamina II was considered a closed system, e.g., composed exclusively of short axon interneurons. According to Ribeiro-da-Silva (1995) such a view is no longer valid, as some cells were found to project to the brain. For example, Lima and Coimbra (1991) claimed that some islet cells project to the reticular formation (RF) of the medulla oblongata. After complex local processing in the DH (Willis and Coggeshall 1991; Parent 1996; Ribeiro-da-Silva 1995) nociceptive signals are conveyed to higher brain centers through projection neurons whose axons form several ascending fiber systems.

Interestingly, after transection of sensory fibers entering the spinal DH or the descending spinal trigeminal tract, the typical substantia gelatinosa-related enzyme acid phosphatase disappeared (Rustioni et al. 1971; Coimbra et al. 1974). Moreover, in the descending spinal trigeminal tract a topographic localization for the ophthalmic, maxillary, and mandibular nerves was described using the disappearance of this enzyme (Rustioni et al. 1971). Later on, fluor-resistant acid phosphatase (FRAP) was related to the nociceptive system (see Csillik et al. 2003).

The central processes of pseudounipolar TG neurons enter the brainstem via the sensory trigeminal root. Some fibers bifurcate to give a rostral branch to the principal (pontine) trigeminal nucleus (PTN) and a caudal branch that joins the spinal trigeminal tract (STrT); some axons only descend to the spinal trigeminal nucleus (STN) (Brodal 1981; Capra and Dessem 1992; Waite and Tracey 1995; Parent 1996; Usunoff et al. 1997; Waite and Ashwell 2004). The PAs terminate somatotopically: most ventral are the ophthalmic fibers, in the middle the maxillary fibers, and dorsally terminate the mandibular fibers. A small number of nociceptive fibers from the 7th, 9th and 10th nerves also join the spinal tract and take a position immediately dorsal to the axons of the mandibular division (Brodal 1947; Usunoff et al. 1997). Generally, the PAs emit collaterals to all three subnuclei of the STN: oralis (STNo), interpolaris (STNi), and caudalis (STNc), defined by Olszewski and Baxter (1954), and according to the classical belief, nociceptive Aδ- and C-fibers terminate almost exclusively in STNc. As suggested at the beginning of the century by Dejerine (1914), inputs from the nose and the lips reach the most rostral parts of STNc, and the posterior regions of the face reach the caudal parts of STNc (onion hypothesis). This appears to be valid from rat to human (Arvidsson 1982; Borsook et al. 2004). Terminations of trigeminal afferents are ipsilateral but some PAs with midline receptive fields terminate in the contralateral STNc (Pfaller and Arvidsson 1988; Jacquin et al. 1990; Marfurt and Rajchert 1991). Many trigeminal PAs reach the paratrigeminal nucleus and solitary nucleus (Usunoff et al. 1997); a moderate number reaches the supratrigeminal nucleus, the dorsal RF, and the cervical SC and a small number of PAs reach cuneate, trigeminal motor, and vestibular nuclei, and even the cerebellum (Marfurt and Rajchert 1991).

The structure of STNc is very similar to the spinal DH (Olszewski and Baxter 1954), and since Gobel et al. (1977) and Gobel (1978a, b), this structure is often called the medullary dorsal horn (MDH) (Craig 1992; Iwata et al. 1992; Li JL et al. 1999; Li YQ et al. 1999, 2000a, b). It has a laminar arrangement with a marginal layer (lamina I), substantia gelatinosa (lamina II), and a magnocellular layer (laminae III, IV). Lamina I is polymorphic, with few large, multipolar neurons (Gobel 1978a; Li YQ et al. 2000a, b), lamina II contains small neurons (Gobel 1978b; Li YQ et al. 1999), and the magnocellular layer actually contains predominantly medium-sized cells, also in humans (Usunoff et al. 1997). In all layers glutamate- and GABA-containing cells are present (Magnusson et al. 1986, 1987; Haring et al. 1990). The GABAergic interneurons innervate the glutamatergic projection neurons, and the latter emit collaterals to the GABA-containing cells (DiFiglia and Aronin 1990). Thus, in the STN there is a reciprocal modulation between the excitatory trigeminothalamic tract (TTT) neurons and the inhibitory interneurons. At the lateral border of the STN, especially in STNc, there are interneurons that immunoreact for NOS (Dohrn et al. 1994; Usunoff et al. 1999). These cells contact the TTT neurons, and Dohrn et al. (1994) suggest that they may indirectly influence orofacial nociceptive processing at the level of the STN via NO production.

In all probability, the MDH is the main, but not the sole part of the trigeminal nuclear complex responsive for nociception. The cornea and the tooth pulp give rise mainly to nociceptive sensations. However, the PAs of these regions reach all components of the trigeminal nuclear complex (Marfurt and Echtenkamp 1988; Barnett et al. 1995; Allen et al. 1996). The rostral parts of the STN also respond to noxious stimulation, and nociceptive responses persist in ventral posteromedial thalamic nucleus (VPM) after trigeminal tractotomy at the obex (Dallel et al. 1988), suggesting nociceptive pathways that are more complex than originally thought (Waite and Tracey 1995).

2.3.1 Types of Terminals in Substantia Gelatinosa

Two types of glomerular terminals could be identified in superficial laminae. One was scalloped, with densely packed clear vesicles of variable size, dark axoplasm, and occasional mitochondria (Figs. 1, 3A,E). These terminals, which contacted several postsynaptic dendrites, correspond to the central terminals of type 1 glomeruli (C1) described by Ribeiro-da-Silva and Coimbra (1982). They are likely to be terminals of unmyelinated PAs (Ribeiro-da-Silva 1995). Terminals of the second type were also scalloped, but with loosely packed clear vesicles of uniform size, light axoplasm and many mitochondria (Figs. 1, 3B,F). These terminals, contacting several postsynaptic profiles and involved in axo-axonic contacts with symmetric active zones, correspond to the central terminals of type 2 glomeruli (C2) described by Ribeiro-da-Silva and Coimbra (1982). These are likely to arise from thinly myelinated PAs (Alvarez et al. 1992, 1993; Light 1992). C1 terminals are concentrated in lamina IIo and dorsal IIi, whereas C2 terminals are concentrated in ventral lamina IIi (Bernardi et al. 1995). Glomeruli make only about 5% of the synapses in substantia gelatinosa (Ralston 1979). The majority of synapses in this region are axo-dendritic, and it is hard to relate them to a particular afferent input. The majority of dome-shaped terminals are believed to originate from intrinsic neurons. Axo-axonic terminals are common in lamina II. Frequently, axo-axonic terminals contain flattened or pleomorphic vesicles (Kerr 1975). Few synapses contain dense core vesicles.

Glutamate Receptors in the Superficial Laminae of the Spinal Cord The superficial laminae of the SC are of particular interest because of their role in hosting the first brain synapse involved in pain processing. This diverse region of the SC also receives other types of PA fibers. Afferents that mediate different types of stimuli (i.e., low- and high-threshold mechanoreceptors) impinge onto the same DH neurons (Willis and Coggeshall 1991). Therefore, the question persists of how spinal neurons decode the convergent inputs at the level of the first synapse. Providing a better understanding about the nature of the synaptic processing in superficial laminae of the SC will directly improve our knowledge and strategies on how to treat abnormal pain. From a pharmacological point of view, a first possibility derives from a speculation that different submodalities are mediated by different neurotransmitters. The pharmacological diversity seems to play a role since the SG neurons giving rise to C-fibers contain substance P, which was not found in cell bodies of normal SG giving rise to A-fibers. Moreover, substance P-positive axons in this area co-localize with μ-opioid receptor (Ding et al. 1995a), suggesting the role of opiates in this region. On the other hand, all PA terminals in the superficial laminae of the SC appear to contain glutamate (Rustioni and Weinberg 1989; Salt and Herrling 1995); nevertheless, the amount of glutamate available in different anatomical classes of terminals may vary (De Biasi and Rustioni 1988; Merighi et al. 1991; Tracey et al. 1991; Levine et al. 1993; Valtschanoff et al. 1994).

In general, a large variety of pre-, post-, and extrasynaptic factors may shape the timing and magnitude of glutamatergic transmission. Normally, glutamate is released by calcium-dependent mechanisms into the synaptic cleft. In the cleft, glutamate is present for brief periods of time because of the fast and highly specific uptake by specific transporters expressed by the nearby astrocytic or neuronal processes and terminals. In the synaptic cleft, glutamate is saturated by two major classes of glutamate receptors: ionotropic and metabotropic. The former are ligand-gated sodium/potassium and, under some circumstances, calcium channels that depolarize the postsynaptic membrane, whereas the latter are coupled to second messenger cascades that can impact metabolism. Three classes of ionotropic glutamate receptors are currently distinct based on their pharmacological characteristics, structure, and physiological properties: AMPA, NMDA, and kainate. AMPA receptors are pore-forming heteromers built-up of a combination of the four subunits: GluR1, GluR2, GluR3, and GluR4. A common property of native AMPA channels is their low affinity to glutamate, blocked by CNQX, and the low permeability of calcium. Local application of CNQX completely abolishes the fast component of the excitatory postsynaptic potentials (EPSP), but does not significantly alter the slower component. Each receptor subunit contributes specific pharmacological and biophysical properties to the receptor channel. For instance, partition of the edited form of the GluR2 subunit into AMPA channels renders them insensitive to internal polyamine block and impermeable to bivalent ions such as calcium.

Different groups of neurons in the brain express a wide variety of AMPA receptor subunit combinations, but not necessarily all of them. Physiological data suggest that this unique phenotyping correlates well with differences in the kinetics of corresponding EPSP. In contrast, NMDA receptors are nonsensitive to CNQX, but to NMDA, show high affinity to glutamate, high voltage dependence due to internal magnesium block, and higher conductance of bivalent ions such as calcium. They are built of an obligatory NMDAR1 subunit and several NMDAR2 subunits. NMDA receptors show lesser variability between brain regions. Finally, kainate receptors have thus far attracted attention particularly because of their presynaptic localization in the superficial laminae of the SC. Their functional significance, at least in the SC, is not clear (Hwang et al. 2001).

Among the number of postsynaptic factors that may contribute to the shape and size of the local glutamatergic depolarization events is the diversity of ionotropic glutamate receptors. Several light microscopic (LM) studies demonstrated high concentrations of AMPA receptor subunits in neurons of superficial laminae of the DH (Furuyama et al. 1993; Henley et al. 1993; Tölle et al. 1993; Tachibana et al. 1994; Kondo et al. 1995; Popratiloff et al. 1996a). However, electron microscopy (EM) was required to verify the presence of receptor subunits at synaptic sites and to explore the relations between receptor subunits and PA terminals. EM evidence for glutamate receptors subunit immunoreactivity was provided with preembedding immunocytochemistry (Liu et al. 1994; Tachibana et al. 1994; Vidnyanszky et al. 1994), suggesting accumulation of electron-dense reaction product at postsynaptic densities. Preembedding was also used in an effort to relate glutamate receptor subunits to PA terminals (Alvarez et al. 1994). Although providing valuable qualitative data, this method was not suitable for quantitative study, both because of variable antibody penetration into the sections and because of the difficulty in quantifying the density of immunoreactions at the EM level. Postembedding immunocytochemistry with colloidal gold can in principle avoid the above technical limitations (Nusser et al. 1995a, b). However, osmic acid used in the classical EM protocols for tissue fixation abolishes or seriously impairs the antigenicity of the vast majority of the proteins, including glutamate receptor subunits. An original method that replaces osmic acid with tannic acid and uranyl salts in material fixed with glutaraldehyde yielded good structural preservation together with precise localization of multiple receptor subunits (Phend et al. 1995). With this technique, relative quantification of AMPA receptor subunits showed that these are highly concentrated at synapses and that functionally different terminals show different affinity to one or another receptor subunit.

Light Microscopic Appearance of AMPA Receptor Subunits in the Substantia Gelatinosa When the immunolabeling was revealed according to a nickel-intensified DAB-peroxidase protocol in two animals, fine granular reaction product in neuronal somata and neuropil was indicative for sites with high concentration of the antigen. Cellular staining could be identified in somata and proximal dendrites. Staining with the GluR1 antibody was concentrated in the superficial DH (Fig. 2A–C). Stained neurons in other regions except lamina X of the SC were small and sparse. Neurons immunoreactive for GluR2/3 were also concentrated in superficial laminae (Fig. 2D–F). However, this antibody also abundantly stained a number of neurons of various size and shape throughout the rest of the SC.

In lamina I, neurons stained with GluR1 were more concentrated laterally (Fig. 2B), whereas a larger population of intensely stained GluR2/3 neurons was present throughout the mediolateral extent of lamina I (Fig. 2E). Fine punctate neuropil staining was present with both antibodies, which was organized in small bundles oriented mediolaterally, especially apparent in the sections labeled with GluR1.

In lamina II, the density of neurons immunostained for GluR1 was highest near the IIo/IIi border; few stained cells were seen in the deep IIi (Fig. 2C). Neuropil staining with GluR1 overlapped the staining of somata, gradually disappearing at the ventral border of lamina II. The staining achieved with GluR2/3 antibody showed a remarkable difference: density of neuronal and neuropil staining was relatively low at the IIo/IIi border, and highest deep in lamina IIi, extending into lamina III (Fig. 2F). GluR2/3 staining is most likely due to the abundance of GluR2 subunit, because the pattern of GluR2 labeling very much resembles those achieved with the GluR2/3 antibody (not shown). Additional results showed that GluR4 antibody produces little and diffuse staining in superficial laminae of the SC. However, recent data suggest that staining with this antibody is concentrated in the presynaptic terminals and these loci are not readily distinguishable with conventional optical microscopy (Lu et al. 2002).

Electron Microscopy With both GluR1 and GluR2/3 antibodies, gold particles were sparse over cell bodies and dendrites. Gold particles were instead clustered over the postsynaptic density, postsynaptic membrane, and cleft of a large number of asymmetric synapses. A large proportion of terminals with positive synaptic zones could be recognized as originating from PAs, together with synaptic zones of many terminals lacking characteristic glomerular organization, likely to originate from intrinsic neurons. Labeling was not observed over active zones of symmetric synapses. Ninety-four percent of gold particles tallied (410/437) from a sample of 215 glomerular terminals from lamina II were in a region between 30 nm outside and 40 nm inside the postsynaptic membrane (Popratiloff et al. 1996a). The majority of gold particles were associated with the postsynaptic membrane and density. The distribution of gold particles was similar for GluR1 and GluR2/3. The very low density of gold particles away from the synaptic active zones implies that even a single gold particle at the active zone is strong evidence for immunopositivity. Examination of serial thin sections confirmed this interpretation, because synapses first identified as labeled by the presence of one gold particle on one section displayed one or more gold particles also in the adjacent sections (Fig. 3C,D). The same did not hold true for gold particles at nonsynaptic sites.

Relationship Between Types of Terminals and Different Receptor Subunits Terminals of both types were presynaptic to both GluR1 and GluR2/3, but to a different extent. C1 synapses were predominantly GluR1-positive, and synapses were predominantly positive for GluR2/3. These differences were highly significant.

Interpretation of the above-mentioned quantitative differences was complicated by the possibility that unlabeled synaptic sites might nonetheless contain receptor subunits, or that the concentration of subunits may vary at different types of synapses. To explore this issue, the number of gold particles underlying each active zone of randomly photographed PA terminals was counted. The counts were roughly Poisson-distributed, reflecting the random exposure of epitopes at the surface of a thin section. However, heterogeneity of synaptic contacts was also suggested, especially for C2 terminals immunopositive for GluR2/3. Immunolabeled C1 synapses contained a similar number of gold particles coding for GluR1, on average, as did immunopositive synapses of C2 terminals (1.88 vs 2.10), confirming that a higher proportion of C2 than of C1 synapses expressed little or no GluR1. On the other hand, immunopositive synapses of C1 terminals contained a markedly lower mean number of gold particles coding for GluR2/3 than did synapses of C2 terminals (1.92 vs 2.79). This could not be explained by differences in dimensions of active zones, because C1 and C2 had active zones of similar lengths (322.6 ± 13 vs 341.6 ± 11 nm, respectively).

Considerations The data on LM distribution of AMPA subunits are generally consistent with previous studies (Furuyama et al. 1993; Henley et al. 1993; Tölle et al. 1993, 1995; Tachibana et al. 1994; Kondo et al. 1995). The high density of AMPA receptor expression in superficial laminae of the DH is consistent with the presence of numerous glutamatergic synapses both from peripheral afferents (Broman et al. 1993; Valtschanoff et al. 1994) and from local interneurons (Rustioni and Cuenod 1982). GluR1-positive neurons are concentrated at the IIo/IIi border and are generally superficial to the GluR2/3-positive neurons. Because previous studies with in situ hybridization suggest that the GluR3 subunit is only weakly expressed in the superficial DH (Furuyama et al. 1993; Henley et al. 1993; Tölle et al. 1993, 1995; Pellegrini-Giampietro et al. 1994), our staining for GluR2/3 is likely to reveal mainly the GluR2 subunit. By extrapolation from observations in the cortex (Kharazia et al. 1996) and in the DCN (Popratiloff et al. 1997), at least a fraction of GluR1-positive neurons in superficial laminae may be GABAergic. Nitric oxide synthase (NOS) coexists with GABA in cells in these laminae (Valtschanoff et al. 1992), and NOS-positive neurons in forebrain lack GluR2 (Catania et al. 1995). However, because NO-synthesizing neurons in the SC are concentrated at the border between laminae II and III (ventral to GluR1-positive neurons), only a modest fraction of GluR1-stained neurons may synthesize NO.

Relationship of LM and EM Results The laminar distribution of staining for the two antibodies was similar at LM and EM. However, staining of somata was prominent at LM, but sparse at EM. This apparent discrepancy is presumably explained by the characteristics of the techniques: immunoperoxidase exhibits high sensitivity (because of Ni-amplification of weak signals by the DAB reaction), but is less well localized than immunogold and does not accurately reflect quantitative differences (Griffiths 1993). Alternatively, the immunogold labeling may require antigen concentration to exceed a threshold value. Craig et al. (1993) provided LM evidence for clustering of AMPA/kainate subunits at synapses in cultured neurons. This was supported by EM immunogold performed on frozen or freeze-substituted sections (Nusser et al. 1994, 1995a, b) and by the present results. The immunoglobulin bridge introduces a localization error of 20 nm for the gold particles (Kellenberger and Hayat 1991). Because staining is confined to the surface, obliquity of synaptic membranes in the section may introduce an additional error of similar magnitude. These errors do not affect the present data concerning the modal location of particles but suggest that our results documenting a strong association of AMPA receptors with the postsynaptic membrane underestimate the precision of this association. The close match between glutamate-enriched terminals and sites immunopositive for glutamate receptors (Craig et al. 1994; Phend et al. 1995) shows that the labeling is selective for excitatory synapses, a conclusion supported by the absence of gold labeling at symmetric synapses.

Number of Receptors at a Synapse The exact numerical relationship between gold particles and receptor molecules cannot yet be determined, but in other systems, one gold particle represents 20–200 molecules of antigen (Kellenberger et al. 1987; Kellenberger and Hayat 1991; Griffiths 1993). This ratio reflects various factors: (a) only antigen molecules presenting an epitope at the surface can be recognized and, even for thin (100-nm) sections, a majority of the epitopes are not exposed; (b) many of the epitopes may be denatured by the fixation and processing; and (c) steric constraints permit only a fraction of surface epitopes to bind immunoglobulin. Thus, although even a single gold particle over a synapse is likely to indicate the presence of a receptor, its absence cannot be taken as proof of the lack of receptor. Nevertheless, because there is an approximately linear relationship between gold particles and antigen density (Ottersen 1989; Griffiths 1993), it is possible to estimate the relative densities of subunits at different synapses. This study is about subunits, not functional receptors. However, considering the high concentration of gold in the vicinity of the postsynaptic membrane, most of these subunits were presumably already in a functionally appropriate position. In cortex and hippocampus, the labeling density seen with this method corresponds well to biophysically derived estimates of functional receptors, assuming a labeling efficiency of 1%–2% (Hestrin 1992; Stern et al. 1992; Griffiths 1993). It can be argued that most subunits inserted into the synaptic membrane have been assembled into functional pentameric receptors.

Relation of Receptors to Types of Synapses C1 terminals contain a low density of mitochondria and a high density of glutamate (Broman et al. 1993; Valtschanoff et al. 1994), both features perhaps related to their lower tonic activity and the need for a larger pool of vesicular glutamate. C1 terminals are frequently presynaptic to GABAergic dendrites, whereas C2 terminals are more frequently postsynaptic to GABAergic profiles, possibly reflecting the generally lower spatiotemporal resolution of unmyelinated vs small myelinated fibers (Bernardi et al. 1995). The present quantitative data show that both types of PA terminals are associated with subtypes of AMPA receptors, but in different proportions. The preference of C1 for GluR1 contrasts with the preference of C2 terminals for GluR2/3 subunits. While the relative role of presynaptic and postsynaptic factors in establishing and maintaining these differences remains to be determined, the contrasting distribution of GluR1 and GluR2/3 immunopositivity raises the possibility that some neurons in the superficial DH may express only one of the two receptor subunits. Because AMPA receptors lacking GluR2 are calcium-permeable (presumably associated with C1 terminals, Hollman and Heinemann 1994), some neurons in the dorsal substantia gelatinosa may experience AMPA-mediated calcium transients in response to glutamatergic synaptic input, particularly that originating from unmyelinated afferents (C1), thus potentially activating second-messenger cascades. Indeed, recent work supported this possibility (Engelmann et al. 1999). Also results from primary culture demonstrate calcium-permeable AMPA channels in some neurons in the DH (Kyrozis et al. 1995). The apparent bias of terminals of unmyelinated fibers toward GluR2-poor AMPA receptors may bear on the issue of hyperalgesia. Sugimoto et al. (1990) proposed that central hyperalgesia secondary to peripheral neuropathy may involve NMDA-mediated excitotoxic damage to inhibitory interneurons. The present data raise the possibility that GABAergic interneurons in substantia gelatinosa may suffer excitotoxic damage from sustained abnormal activity in unmyelinated fibers synapsing onto calcium-permeable AMPA channels.

NMDAR1 and Primary Afferent Terminals in the Superficial SC With the nickel-intensified DAB-peroxidase procedure, immunostaining at the LM level produced a fine granular product in cells and neuropil. In 25-µm sections, cellular staining could be identified in somata and proximal dendritic arbors. Within the DH, staining was more prominent in the superficial laminae, especially lamina II, possibly because of its higher cellular concentration (Fig. 4A,B). Neuropil staining was densest in lamina I and IIo and tended to decrease more ventrally in the superficial dorsal horn (Fig. 4B). This was confirmed in plastic embedded, 1-µm-thick sections in which staining was denser in IIo where cells are more densely packed (Popratiloff et al. 1998b).

At the EM level, sections showed generally good structural preservation in the absence of osmium fixation (see also Feirabend et al. 1994, 1998). Myelin was poorly preserved but clear, and dense core vesicles as well as synaptic specializations were well preserved and contrasted. Gold particles were sparse over cell bodies and dendrites but more frequently encountered than in sections stained for AMPA receptors. Particles were clustered over the postsynaptic density, pre- and postsynaptic membrane, and over clefts of a large number of asymmetrical synapses. A significant fraction of terminals with positive synaptic zones could be recognized as originating from primary afferents, but synaptic zones of many terminals of uncertain origin were also immunopositive. Labeling was not observed over active zones of symmetric synapses. In addition to scalloped terminals at the center of C1 (Fig. 4C,D) and C2 (Fig. 4E) glomeruli, a third distinct group of terminals in superficial laminae are dome-shaped. They display loosely packed clear vesicles of irregular size, light axoplasm, and many dense core vesicles (DT in Fig. 1). These terminals are not involved in glomerular arrangement and contact, in the plane of transverse ultrathin section, only a single dendrite or dendritic spine. They are concentrated in lamina I, extending into IIo. Many of these terminals are of primary afferent origin.

To explore whether there is a different concentration of the receptor subunit at different classes of terminals, gold particles underlying active zones were counted for each group of terminals from random photographs. As expected, the counts were roughly Poisson-distributed, reflecting the random exposure of epitopes in a thin section. Immunopositive C1 (Fig. 4C,D) and C2 (Fig. 4E) terminals had similar counts of gold particles (2.18 ± 0.13 and 2.06 ± 0.13, respectively) and these were lower than the counts for nonglomerular terminals (2.36 ± 0.17). This difference is likely to be explained by differences in the length of active zones between glomerular and nonglomerular terminals, i.e., on one side 266 ± 26 for C1 terminals and 268 ± 18 nm for C2 terminals, respectively, and on the other side 387 ± 24 nm for nonglomerular terminals.

The apparently uniform relationship between NR1 sites and the three types of terminals considered here differs from the results of a study with AMPA subunits (Popratiloff et al. 1996a). Additional data show also that nonglomerular terminals contact postsynaptic sites with GluR2/3 subunits about twice as frequently as postsynaptic sites with the GluR1 subunit. These data show that most PA synapses in superficial laminae express NR1; considering the limited sensitivity of immunogold. These data are also compatible with the expression of NMDA receptors at all such PA synapses. Available data generally support that, as for other regions of the CNS, synaptic potentiation requires activation of NMDA receptors, though it may be expressed mainly via AMPA receptors. The present data thus suggest that virtually all primary afferent synapses in the superficial DH may be potentiated, although in view of previously reported results, this may further strengthen expression of different AMPA subunits for different groups of synapses.

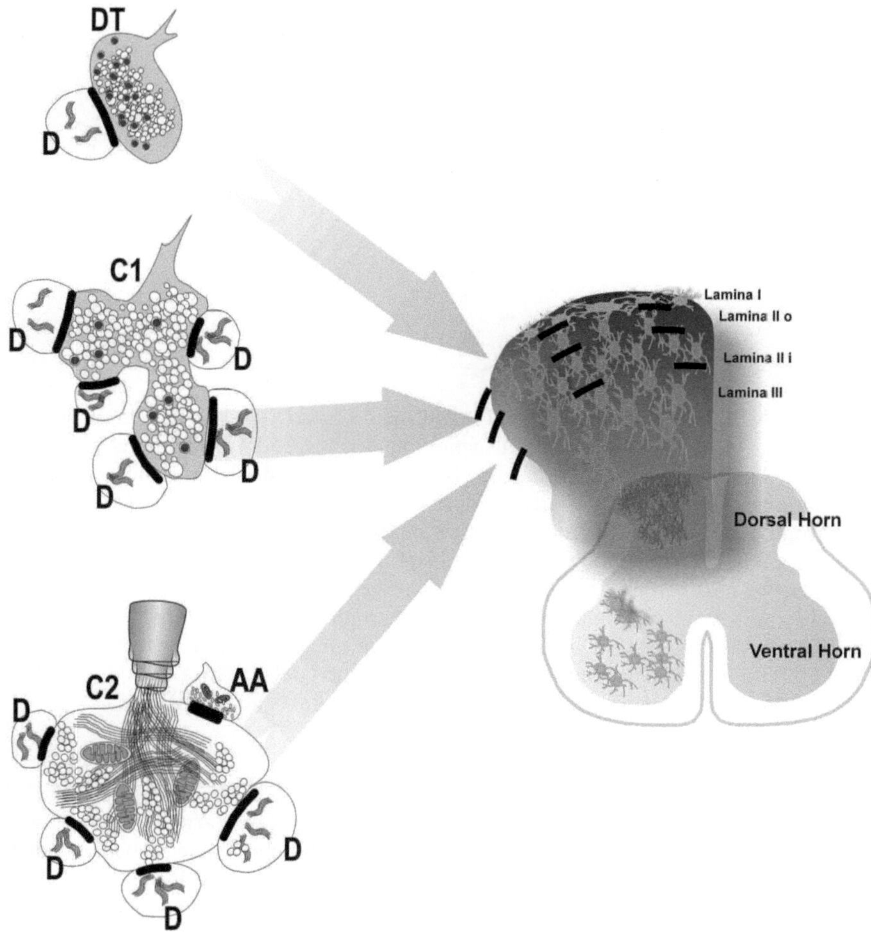

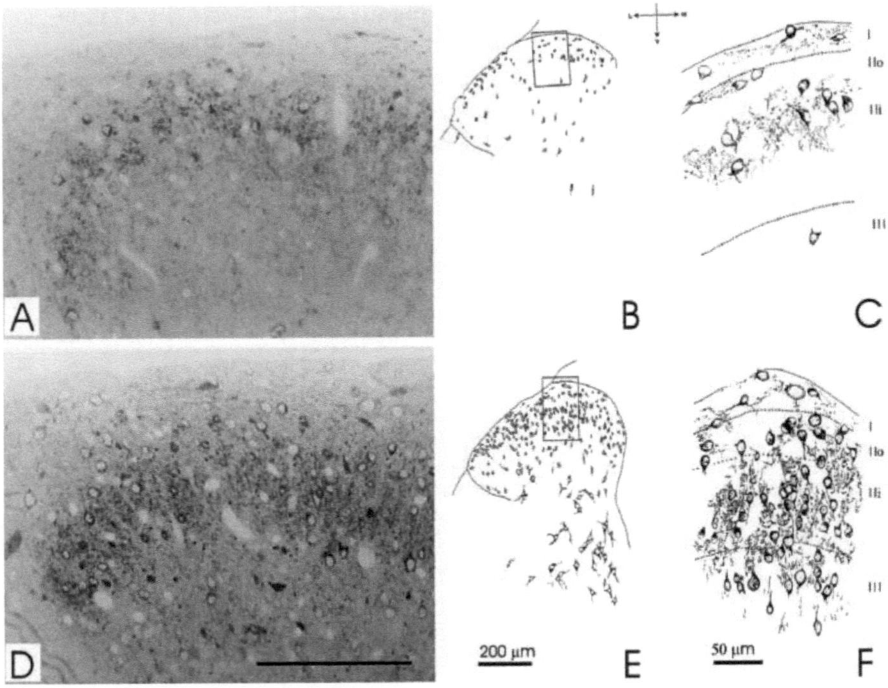

Fig. 2 A–F AMPA receptor subunits GluR1 (**A–C**) and GluR2/3 (**D–F**) in the rat substantia gelatinosa. **A** An image from semithin section labeled for GluR1. Labeling is present in neuronal cell bodies and neuropil. Labeling is denser at the border between outer lamina II (*IIo*) and inner lamina II (*IIi*), whereas in deep lamina IIi it is present as sparse punctae in the neuropil. **B** Low-power camera lucida drawing from a 50-µm-thick section labeled with GluR1 antibody, and **C** high power from the box on **B**, showing differential density of the GluR1 labeling in superficial laminae (*I–III*) of the DH. **D–F** In contrast to GluR1, GluR2/3 labeling is present in neuronal perikarya and neuropil through laminae I–III. Staining density increases from lamina I to lamina III. **D** A semithin section similar to **A** labeled for GluR2/3; **E** and **F** camera lucida drawings similar to **B** and **C** labeled for GluR2/3. Scale bar: **D** and **A**, 200 µm. (Adapted with permission from Popratiloff et al. 1996a)

Fig. 1 Schematic drawing representing the three major types of primary afferent terminals that could be distinguished by their morphology. *Upper left*, small dome shaped terminals (*DT*), which contain a few large dense core vesicles and contact a single dendrite (*D*). These terminals are more abundant in lamina I. *Central left*, a large scalloped terminal at the center of type 1 glomerulus (*C1*). These terminals have dark axoplasm, densely packed vesicles of various sizes and occasional large dense core vesicles. C1 terminals contact several dendrites and are more abundant in lamina IIo. *Bottom left*, large scalloped terminal at the center of type II glomerulus (*C2*). The terminals contain sparse clear vesicles, many neurofilaments and several mitochondria. Such terminals also contact several dendrites, but are more frequently postsynaptic to inhibitory axo-axonic terminals (*AA*). These terminals are concentrated in laminae IIi and III

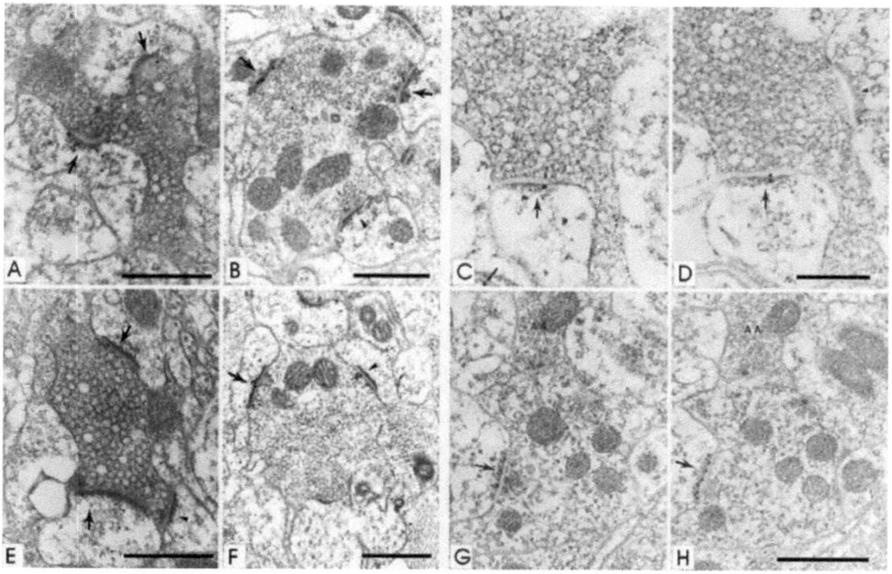

Fig. 3 A–H AMPA receptor subunits GluR1 (**A–D**) and GluR2/3 (**E–H**) at the central terminals of C1 (**A, C, D, E**) or C2 (**B, F, G, H**) in the substantia gelatinosa of the rat DH revealed with postembedding immunogold. More frequently active zones of C1 (**A, C, D,** *arrows*) than C2 (**B,** *arrows*) terminals were labeled for GluR1. However, strongly labeled active zones were present at both C1 (**A,** *left arrow*) and C2 terminals (**B,** *left arrow*). In contrast, GluR2/3 more frequently labeled terminals of C2 (**F, G, H**) than C1 (**E**) glomeruli. On average more gold particles were found at C2 active zones, compared to C1. **C, D** Serial sections through a same C1 terminal labeled with GluR1, and **G, H** serial sections through a same C2 terminal labeled with GluR2/3. *Arrows* show positive active zones, *arrowheads* (**B, D, E, F**) point to negative active zones. *AA* axo-axonic terminal. Scale bars: **A, B, E, F, G, H,** 500 nm; **D** and **C,** 250 nm. (Adapted with permission from Popratiloff et al. 1997)

Fig. 4 A Low-power camera lucida drawings from a 50-μm-thick vibratome section stained with anti-NMDAR1 antibody. **B** Higher-power camera lucida drawing from the field in box on **A.** NMDAR1 antibody stained uniformly perikarya and neuropil through laminae I–III. **C–E** NMDAR1 immunolabeling detected with postembedding immunogold in C1 **C, D** and C2 **E** PA terminals. Gold particles labeling was weaker than those observed for AMPA receptor subunits. **C, D** Consistently low labeling in serial sections through a same C1 terminal (*arrow,* positive active zone; *arrowhead,* negative active zone). **E** NMDAR1 antibody stains weakly the active zones of C2 PA terminals (*arrow*), some gold particles are present presynaptic (*arrowhead*), and a few active zones show accumulation of more than two gold particles (*open arrow*). Note that the symmetric contacts are negative for NMDAR1 (*thick arrow*). Scale bars: **A,** 200 μm; **B,** 50 μm; **D** and **C,** 250 nm; **E,** 500 nm. (Adapted with permission from Popratiloff et al. 1998b)

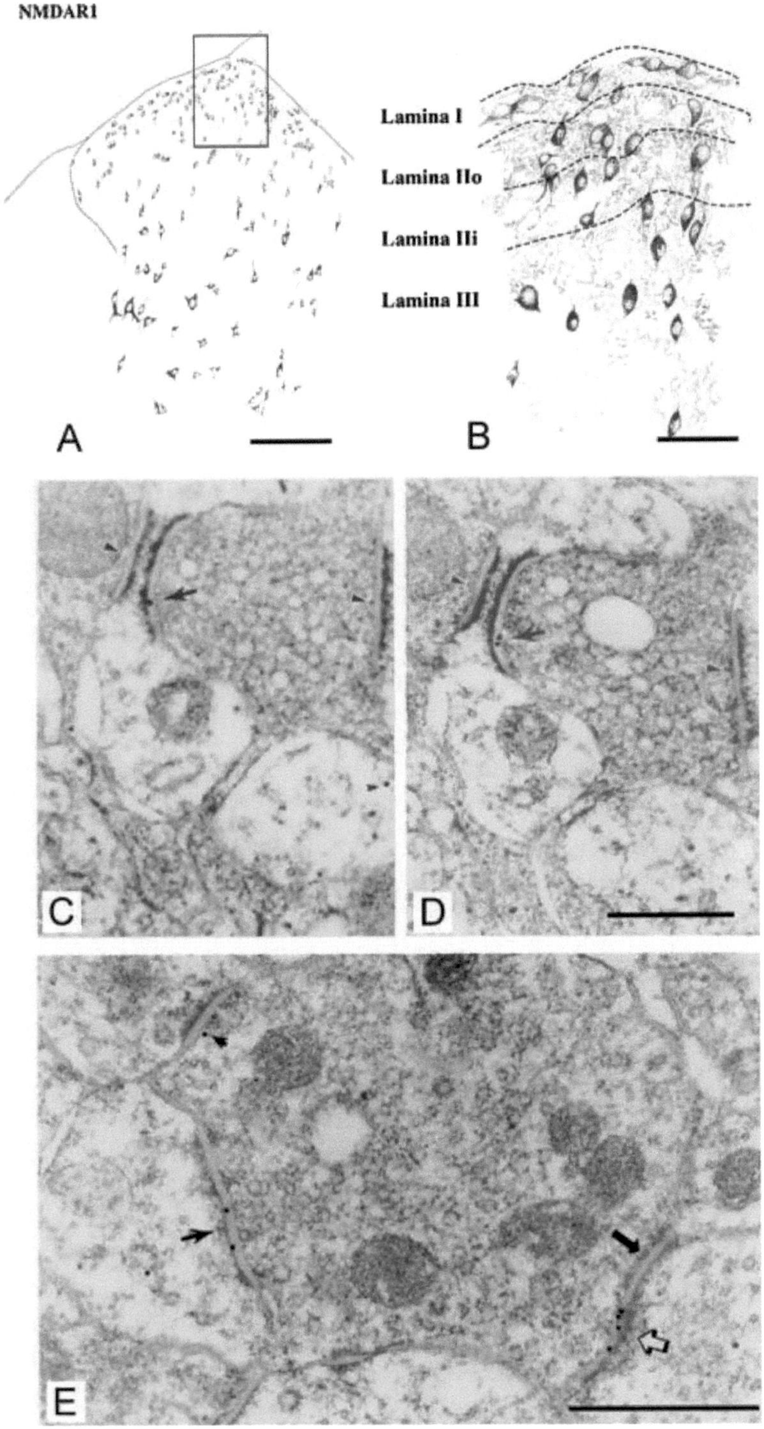
NMDAR1
Lamina I
Lamina IIo
Lamina IIi
Lamina III
A
B
C
D
E

2.4
Ascending Pathways of the Spinal Cord and of the STN

2.4.1
Spinothalamic Tract

In experimental animals, it was repeatedly reported that the large lamina I neurons are the source of about one-half of the spinothalamic tract (STT) (Willis et al. 1979; Apkarian and Hodge 1989a, b; Craig 1995). Recently, however, Klop et al. (2004a, b) declared that in the cat the percentage of lamina I neurons is 4.9%–14.2% of the 12,000 spinothalamic neurons in the SC. The STT in humans mediates the sensations of pain, cold, warmth, and touch (Hassler 1960; Kerr 1975a; Nathan and Smith 1979; Brodal 1981; Jones 1985, 1998; Willis 1985; Willis and Coggeshall 1991; Craig 1996a; Willis and Westlund 1997, 2004; Nathan et al. 2001). The mean conduction velocity of the STT estimated in experimental animals is approximately 8.0 m/s (Dostrovsky and Craig 1996). The modern physiological methods also allow its evaluation in humans (Rossi et al. 2000; Tran et al. 2002). The mean conduction velocity was estimated by Rossi et al. (2000) to be approximately 9.87 m/s.

The cells of origin are located mainly in laminae I and IV–VI. Few STT neurons are located in lamina X (around the central canal), and in laminae VII and VIII (in the ventral horn, dorsal to the "motoneuronal" lamina IX) (Willis et al. 1979; Granum 1986; Kemplay and Webster 1986; Apkarian and Hodge 1989a, b; Burstein et al. 1990b; Willis and Coggeshall 1991; Craig 1996b; Usunoff et al. 1999; Andrew and Craig 2001). The neurotransmitter of the STT neurons is glutamate (Ericson et al. 1995; Blomqvist et al. 1996) and the STT cells also express peptides as co-transmitters (Ju et al. 1987; Battaglia et al. 1992; Battaglia and Rustioni 1992; Todd and Spike 1993; Broman 1994). Lee et al. (1993) claimed that some STT neurons contain NOS, but for the contrary see Kayalioglu et al. (1999) and Usunoff et al. (1999). Most of the cells project to the contralateral thalamus. However, in experimental animals a fairly significant number of ipsilaterally projecting cells (approximately 10% of the total STT neuronal population) were detected (Burstein et al. 1990b). Clinical observations indicate that ipsilaterally projecting STT neurons also exist in humans (Nathan et al. 2001). The STT axons cross the midline in the commissura alba anterior transversely, rather than diagonally (Nathan et al. 2001) and ascend in the anterolateral quadrant of the SC white matter. The axons of lamina I neurons in the monkey ascend more dorsally than do the axons of neurons in the deeper laminae (Apkarian and Hodge 1989b), and in the cat the ascending fibers of the lamina I cells are scattered throughout the lateral white matter (Craig 1991). Clinical evidence from anterolateral cordotomies in patients with intractable pain indicates that the STT axons are somatotopically arranged. The axons representing the lower extremity and the caudal body parts are located more laterally, and those representing the upper extremity and the cranial body parts more anteromedially (Nathan and Smith 1979; Lahuerta et al. 1994; but see Marani and Schoen 2005 for debate). In the brainstem, the STT ascends close to the dorsolateral wedge of the medial lemniscus (Walker 1940; Bowsher 1957; Hassler

1960; Mehler et al. 1960; Mehler 1962). The axons that reach the thalamus are very few in number. In all probability, a large amount of fibers end in the brainstem. The STT starts in the spinal cord with over 10,000 axons. Glees and Bailey (1951) and Bowsher (1963) counted in the rostral midbrain approximately 1,000 axons with diameters of 2–4 μm, and only 500 axons with diameters of 4–6 μm, and the area occupied was only 0.8 mm in width. In humans and primates, the STT axons terminate in the caudal and oral parts of the nucl. ventralis posterior lateralis (VPLc and VPLo), the nucl. ventralis posterior inferior (VPI), the medial part of the posterior nuclear complex (Pom), nucl. centralis lateralis (CL), as well as in other intralaminar and medial nuclei (Walker 1940; Hassler 1960; Mehler 1966; Kerr 1975b; Boivie 1979; Mantyh 1983; Apkarian and Hodge 1989c; Cliffer et al. 1991; Ralston and Ralston 1992, 1994; Willis et al. 2001, 2003; for the delineation of the thalamic nuclei see Hassler 1959, 1982; Jones 1985, 1997a, b, 1998; Hirai and Jones 1989; Mai et al. 1997; Ralston 2003; Percheron 2004; Marani and Schoen 2005).

There is a large body of literature on the STT in subprimate species (Lund and Webster 1967b; Carstens and Trevino 1978a, b; Willis et al. 1978, 1979; Giesler et al. 1979, 1981; Kevetter and Willis 1982, 1983, 1984; Peschanski et al. 1983; Granum 1986; Craig 1987, 1991, 1995, 2003b, d; Lima and Coimbra 1988; Stevens et al. 1989; Burstein et al. 1990b; Cliffer et al. 1991; Tracey 1995; Shaw and Mitrofanis 2001; Andrew and Craig 2002; Gauriau and Bernard 2004; Klop et al. 2004a, b), but it should be interpreted with caution, since the organization of STT and thalamocortical projections related to pain is fundamentally different in primate species than in nonprimate species such as rodents and carnivores (Craig and Dostrovsky 1999; Blomqvist and Craig 2000; Marani and Schoen 2005). Percheron (2004) pointed out that there are also noticeable changes from monkeys to man: thalamic parts have disappeared, others have appeared, and some have considerably developed (see also Marani and Schoen 2005). In the cat, lamina I STT axons terminate in nucl. submedius, a significant relay nucleus for nociception (Craig 1987; Ericson et al. 1996). Craig et al. (1994) defined in the monkey clusters of nociceptive and thermoreceptive specific neurons, reached by lamina I STT axons, located in the posterior part of the nucl. ventralis medialis (VMpo). Blomqvist et al. (2000) identified VMpo also in the human thalamus; it is included in the suprageniculate/posterior complex of Hirai and Jones (1989), and corresponds to the nucleus limitans portae (located immediately caudal to the nucl. ventrocaudalis parvocellularis internus), and adjacent part of nucleus ventrocaudalis portae of Hassler (1960, 1982). The VMpo is proportionally much larger in humans than in monkeys (Blomqvist et al. 2000) and coincides with the dense zone of STT input recognized by Mehler (1966) in human posterolateral thalamus (Lenz et al. 2000). The proposal of Blomqvist et al. (2002) that STT axons do not terminate in VPL was reviewed by Willis et al. (2001, 2002). Also, Graziano and Jones (2004) questioned the existence of VMpo as an independent thalamic pain nucleus or as a specific relay in the ascending pain system in the monkey. According to Craig et al. (1994) and Craig (1998, 2000), lamina I in primates projects to three thalamic zones: (a) VMpo, (b) VPI, which receives convergent input from lamina V and the dorsal column nuclei, and (c) to a small zone in the medial thalamus (MDvc), which receives a STT input predominantly from lamina I. The VMpo projects topographically to the fundus of the superior limiting sulcus of the insular cortex and to area 3a in the fundus of the central sulcus (Craig 1996a, 2000). MDvc projects to the fundus of the anterior cingulate cortex (field 24c) (Craig 2000). Interestingly, the termination of STT axons in the lateral habenular nucleus escaped recognition, and was only recently described by Craig (2003b) as arising in lamina I in the cat. According to Craig (2003b), the spinohabenular connection could be significant for homeostatic behaviors.

The dorsal column nuclei (DCN), consisting of nucleus gracilis (Gr) and nucleus cuneatus (Cu) are traditionally regarded as a structure primarily involved in conscious fine tactile sensation. The basis for this designation is the DCN's well-established role in relaying precise tactile information from primary dorsal column fibers to the VPL and from there to the somatosensory cortex. However, there is growing evidence that the DCN are also strongly involved in nociception. The DCN project via the medial lemniscus to VPL, Po, and zona incerta, as well as to the border zone between VPL and VL (Lund and Webster 1967a; Boivie 1978; Berkley et al. 1980, 1986; Peschanski and Ralston 1985; Kemplay and Webster 1989; Marani and Schoen 2005). The DCN-thalamic projection is glutamatergic (De Biasi et al. 1994). The connection is constantly described as completely crossed, and only Kemplay and Webster (1989) mentioned occasional ipsilaterally projecting neurons. According to Wree et al. (2005), however, about 5% of the DCN neurons project to the ipsilateral thalamus in the rat.

Ralston and Ralston (1994) compared the mode of termination of STT and medial lemniscal axons and found that the thalamic synaptic relationships of these two thalamopetal systems are fundamentally different. The terminals of the medial lemniscus very often contact (46% of the synaptic contacts) the GABAergic interneurons, which in turn contact the relay neurons. In contrast, more than 85% of the spinothalamic afferents form axodendritic synapses with relay cells, and only in 4% the STT terminals contact the GABA-immunoreactive presynaptic dendrites. Ralston and Ralston (1994) pointed out that because the STT neurons predominantly transmit information about noxious stimuli, the simple axodendritic circuitry of the majority of these spinal afferents suggests that the transmission of noxious information is probably not subject to GABAergic modulation by thalamic interneurons, in contrast to the GABAergic processing of non-noxious information carried out by the medial lemniscus afferents. On another hand, Ericson et al. (1996) found that the lamina I terminations in the nucleus submedius of the cat also participate in synaptic triads, synapsing on presynaptic vesicle-containing dendrites of the interneurons. Beggs et al. (2003) investigated the termination of lamina I STT axons in VMpo in macaques. They reported that these synaptic boutons are relatively large and contain densely packed, round synaptic vesicles. The STT terminals make asymmetric synaptic contacts on low-order thalamic neurons. Similar to Ericson et al. (1996), Beggs et al. (2003) found that the STT terminals are closely associated with GABAergic presynaptic dendrites, and nearly all form classic triadic arrangements (axo-dendro-dendritic synapse).

The critical role of the STT in pain is universally acknowledged, but the relative involvement in pain sensation of lamina I neurons and the wide-dynamic-range lamina V neurons is controversial (Willis and Westlund 1997; Price et al. 2003 vs Craig 2004). According to Price et al. (2003) the wide-dynamic-range lamina V STT neurons are necessary and sufficient for all types of pain sensation and their discharge encodes pain. On the other hand, Craig (2004) reported that, in the monkey, the burning pain is signaled by modality-selective lamina I neurons and not convergent lamina V wide-dynamic-range STT cells.

Primate STT neurons that project to the lateral thalamus (VPL) have receptive fields on a restricted area. Therefore, they are well suited to a function in signaling the sensory-discriminative aspects of pain (Willis et al. 1974; Willis and Westlund 1997, 2004). Primate STT cells that project to the CL may also collateralize to the lateral thalamus, and have response properties identical to those STT neurons that project just to the lateral thalamus (Giesler et al. 1981; Willis and Westlund 1997). On the other hand, STT neurons that project only to the CL have very large receptive fields (Giesler et al. 1981; Willis and Westlund 1997).

The entire trigeminal sensory nuclear complex projects to the thalamus (Peschanski 1984; Magnusson et al. 1987; Mantle-St. John and Tracey 1987; Jacquin et al. 1989; Kemplay and Webster 1989; Dado and Giesler 1990; DiFiglia and Aronin 1990; Iwata et al. 1992; Williams et al. 1994; Barnett et al. 1995; Waite and Tracey 1995; Usunoff et al. 1997, 1999; Li 1999; Li JL et al. 1999; Zhang and Yang 1999; Hirata et al. 2000; Graziano and Jones 2004). The trigeminothalamic tract (TTT) projections are not uniform. Following unilateral horseradish peroxidase injections into the thalamus, Kemplay and Webster (1989) counted 8,683 retrogradely labeled neurons in the PTN, 524 cells in the STNo, 1,920 neurons in the STNi, and 260 labeled cells in the STNc. Generally, the projection toward the VPM and the posterior thalamic nucleus (Po) arises mainly in PTN and in STNi, while the nucl. submedius and the intralaminar nuclei are heavily innervated by the nociceptive STNc. The lamina I neurons send strong projections to the nucl. submedius, VPM, and Po. The deeper laminae moderately innervate VPM and Po, but project heavily to the ventral diencephalon (see the following section). The smallest thalamic innervation (to VPM and Po) arises in STNo. The TTT is bilateral but, especially for the STN, strongly crossed.

2.4.2 Projections to the Ventrobasal Thalamus in the Rat

We examined the projections of the trigeminal sensory nuclei, DCN, and the SC to the thalamus by means of the retrograde axonal transport fluorescent method of Kuypers et al. (1980). We injected unilaterally in the thalamus of Wistar rats ($n = 20$) 2 μl of 1% Fast Blue (FB, Sigma, dissolved in physiological saline), 0.5 μl per injection focus (Fig. 5). Two injections were placed 6 mm, and two 5 mm anterior to the interaural line. The injection foci spread to all somatosensory thalamic nuclei on the side of the injection, including the ventrobasal complex (VPL and VPM), posterior nucleus group, and the intralaminar nuclei. Animals were transcardially perfusion fixed 5 days after injection. This fluorescent dye labels the cytoplasm silver blue, and in heavily loaded cells extends also in the dendrites. The FB injection foci are sharply demarcated (Fig. 5), and it is successfully transported over long distances. The present results are comparable with our previous data, obtained with a very effective retrograde tracer colloidal gold conjugated to the B subunit of cholera toxin (Usunoff et al. 1999).

In the brainstem, the PTN and the three subdivisions of the STN contained retrogradely labeled neurons, but to a very different extent (Figs. 6–9). The largest number of retrogradely labeled neurons was observed in the PTN, contralateral to the injection. From its rostral to its caudal pole, this nucleus was filled with densely packed labeled neurons that formed vaguely delineated clusters (Fig. 6A). The ipsilateral PTN contained a moderate number of FB-labeled neurons, mainly in its dorsal sector (Fig. 6B). The TTT neurons are multipolar, rarely exceeding 20 µm. In the ventral part of the PTN, the neurons are slightly larger. In the STNo, the labeling sharply decreases (Fig. 7). Throughout the entire rostrocaudal extent of STNo, the labeled neurons were more concentrated in the ventral part of the nucleus. The cells are slightly smaller than in the PTN, usually about 18 µm, but some neurons measure about 30 µm (Fig. 7A). There were also few ipsilaterally projecting neurons (Fig. 7B), and most of these cells measured less than 18 µm. The contralateral STNi contained a substantial number of FB-labeled neurons (Fig. 8A,B). Especially in more caudal sectors, some features of lamination were seen (Fig. 8B). The labeled neurons vary considerably in size and shape: from small, rounded to larger, heavily loaded with FB multipolar perikarya. We observed only occasional ipsilaterally projecting neurons in STNi. Throughout the contralateral STNc the retrograde labeling was moderate (Fig. 9). Toward the spinomedullary junction, the number of FB-labeled neurons gradually decreased. Most laterally in the STNc were the characteristic marginal cells (medullary lamina I) (Fig. 9A). They were usually elongated and oriented parallel to the spinal trigeminal tract. Within the latter also few labeled neurons were seen (Fig. 9A). Few labeled neurons were seen in the magnocellular layer (laminae III, IV). Actually the cells were medium-sized, with average diameters of about 20 µm. The ipsilateral projection to the thalamus from the STNc is faint but unquestionable. Almost exclusively marginal neurons were labeled (Fig. 9B).

The present experiments demonstrate a prominent crossed connection from the DCN to the thalamus, from the rostral (Fig. 10A) to the caudal pole (Fig. 10B) of the nuclear complex. The FB-labeled neurons are medium to small in size, measuring approximately 20 µm. Few neurons in the DCN ipsilateral to the injection were labeled (Fig. 10A,B), mostly one to three per section.

For cytoarchitectonic orientation in the SC, the atlas of Molander and Grant (1995) was consulted. The distribution of labeled neurons was very uneven. The highest number of STT neurons was encountered at the spinomedullary junction (Fig. 11), and in the four cranial cervical segments (C1–C4), contralateral to the thalamic injection (Fig. 12). At these levels, a prominent cell labeling was also observed in the lateral cervical nucleus (LCN) (Figs. 11, 12). Notably, also in the first four cervical segments, there was only a moderate number of labeled marginal, lamina I neurons. Most significant labeling was found near the medial aspect of the DH, in the medial extension of lamina IV and adjacent lamina V. Scattered labeled neurons were observed in laminae V–VIII. The ipsilateral STT arising in the first four segments is substantial. Most of these neurons are located deep in the ventral horn, lamina VIII, adjacent to the motoneuronal lamina IX (Fig. 12) Only a few lamina I cells project to the ipsilateral thalamus (Fig. 12). Starting from the fifth cervical segment, the number of STT neurons sharply diminishes (Fig. 13). Very few cells were seen in lamina I, and there were few in the deeper lamina of the DH. Occasional labeled neurons were seen in lamina (area) X (Fig. 13). The ipsilateral STT from the lower cervical segments was very scant.

The thoracic SC of the rat contained only few STT neurons, especially in the cranial thoracic segments (Figs. 14, 15). Singly scattered cells were seen in lamina I, in the deeper laminae, as well as in lamina X. Although very few, ipsilaterally projecting neurons were seen (Fig. 14).

In the lumbar segments, the number of STT neurons increased (Figs. 16, 17). Labeled neurons in lamina I were very few. Scant FB-labeled neurons were seen in the lateral spinal nucleus (LSN) (Fig. 16A). More numerous were the cells in the deeper laminae (Fig. 16B), as well as in area X (Fig. 17). Some larger cells were heavily labeled and FB extended also into the dendrites. Although few, ipsilaterally projecting STT neurons were also present (Fig. 16B).

In the sacral (Fig. 18A,B) and coccygeal (Fig. 18C,D) segments only few, but heavily labeled neurons were found in the DH. STT neurons in lamina I were practically absent, but few such were seen in the LSN, and in this structure were located the occasional ipsilaterally projecting cells. Most STT neurons were found in the deep laminae of the DH, in area X, and in the dorsal laminae of the ventral horn. Some neurons are heavily loaded with FB and occasionally one was able to follow the retrogradely labeled axon (Fig. 18A,B).

Fig. 5 (top) Low-power photograph of the maximum extent of the two rostral injection foci. By the medial focus, also the distal part of the needle tract is filled with Fast Blue. The four injection foci fused ventrally and completely engaged VPL and VPM, as well as considerable portions of Po, and the intralaminar nuclei. To the *lower left*, the dorsal part of the third ventricle (*III*). Despite the massive injection, there is no spillage of FB to the contralateral side, so that the findings below on the ipsilateral TTT and STT, as well as for the DCN-thalamic projection are reliable. Scale bar: 200 μm

Fig. 6 A (bottom) The contralateral principal trigeminal nucleus (*PTN*) is filled with regularly packed retrogradely labeled neurons, **B** while a few such cells in the ipsilateral PTN are concentrated in its dorsal part. For orientation, the laterally adjoining spinal trigeminal tract (*STrT*) is indicated. Scale bars: 100 μm

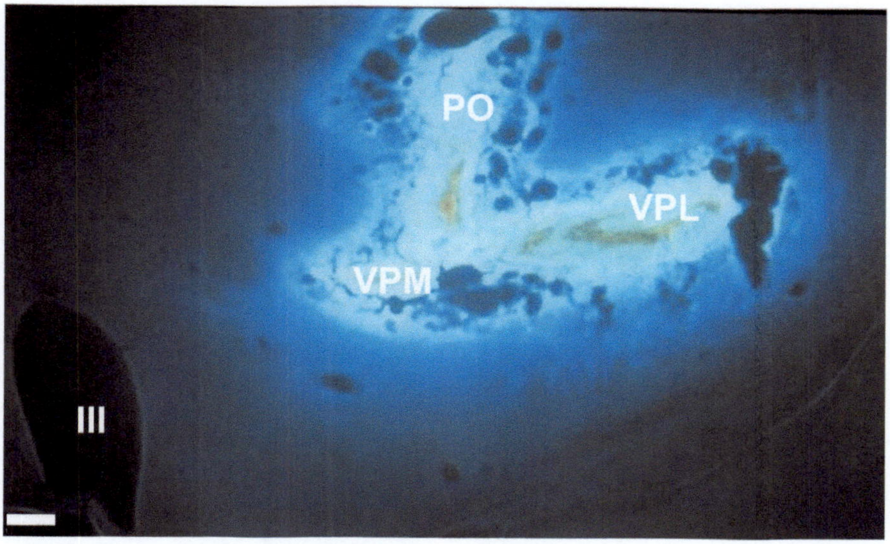

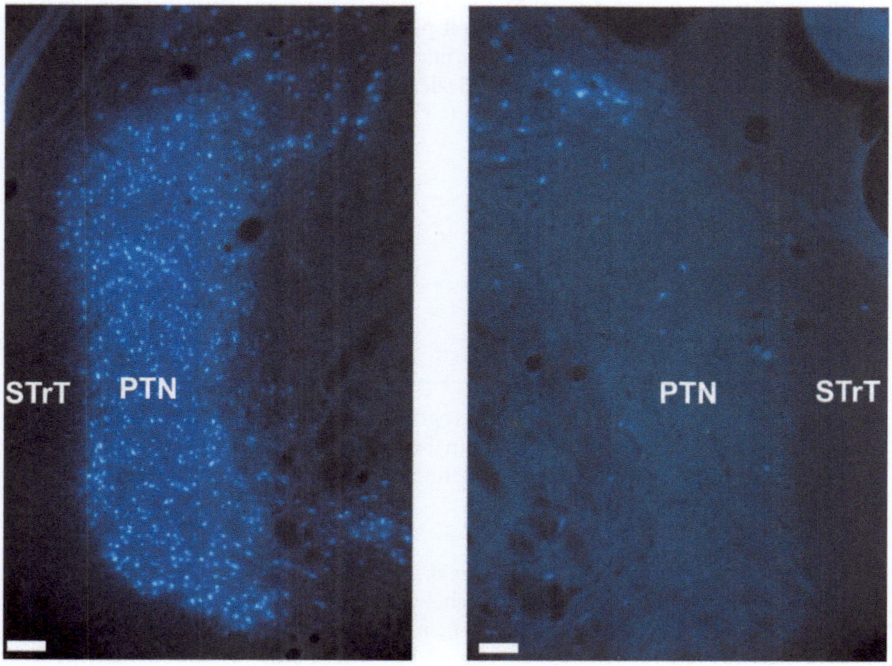

A

B

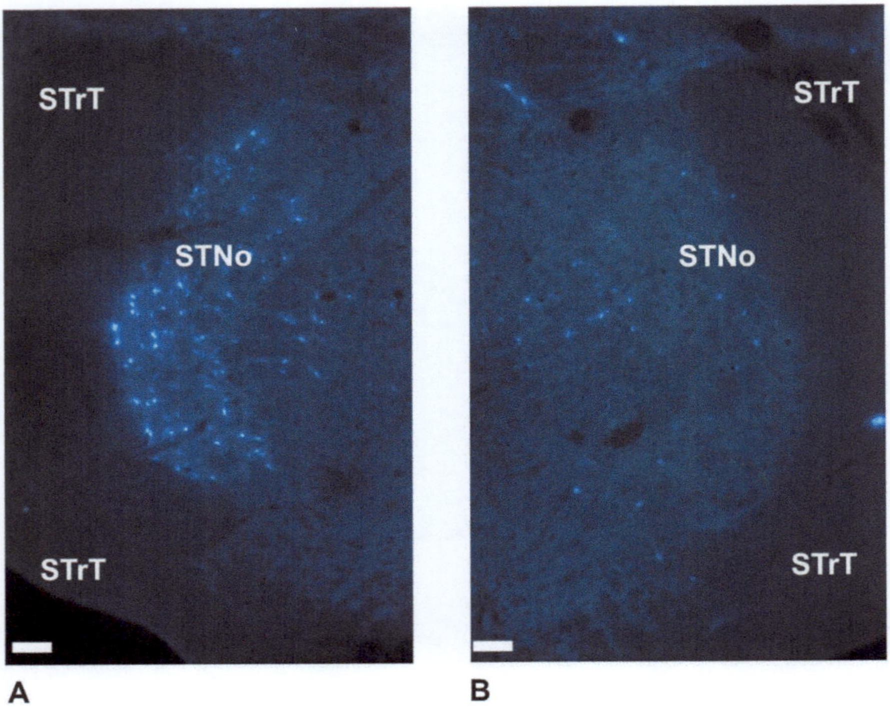

Fig. 7 A In the contralateral spinal trigeminal nucleus, oral part (*STNo*) retrogradely labeled neurons preferably are located in its ventral part, **B** while a few labeled neurons in the ipsilateral STNo are scattered throughout the nucleus. For orientation the laterally adjoining spinal trigeminal tract (*STrT*) is indicated. Scale bars: 100 μm

Fig. 8 A,B (top) A significant number of retrogradely labeled neurons are homogeneously distributed throughout the contralateral spinal trigeminal nucleus, interpolar part (*STNi*), both **A** rostrally and **B** caudally. For orientation, the laterally adjoining spinal trigeminal tract (*STrT*) is indicated. Scale bars: 100 μm

Fig. 9 A,B (bottom) Compared with the large mass of the spinal trigeminal nucleus, caudal part (*STNc*) the number of retrogradely labeled neurons in the contralateral nucleus is relatively low (**A**). There are several labeled neurons also seen in the STrT. In the ipsilateral STNc, labeled neurons are observed in lamina I, just at the border with the STrT. Scale bars: 100 μm

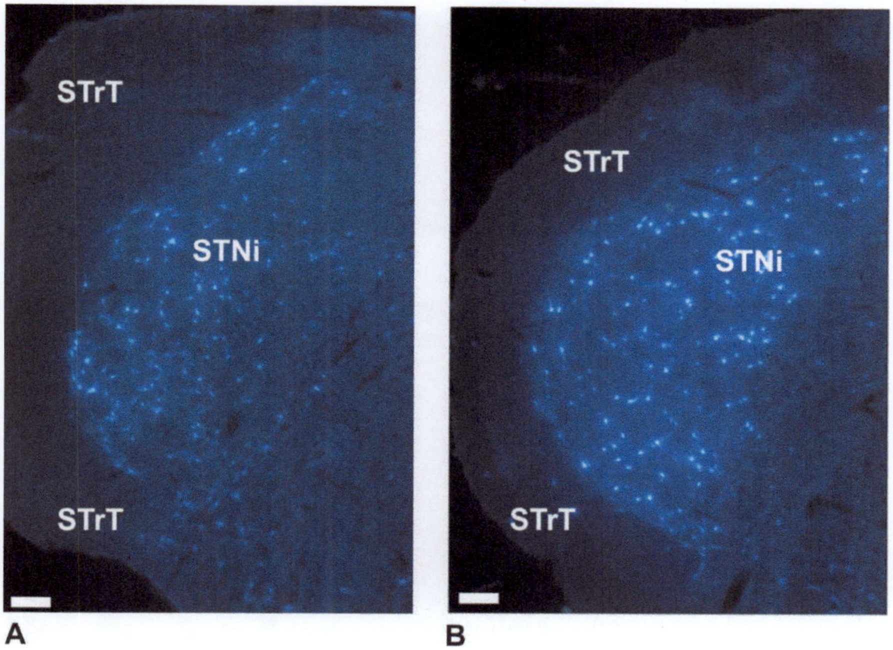
STrT
STNi
STrT
A
STrT
STNi
STrT
B

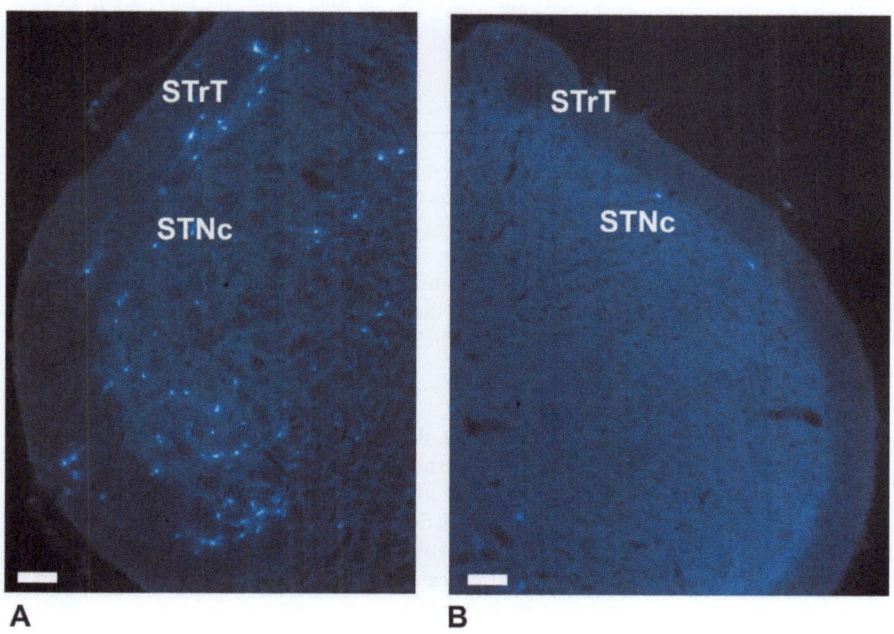
STrT
STNc
A
STrT
STNc
B

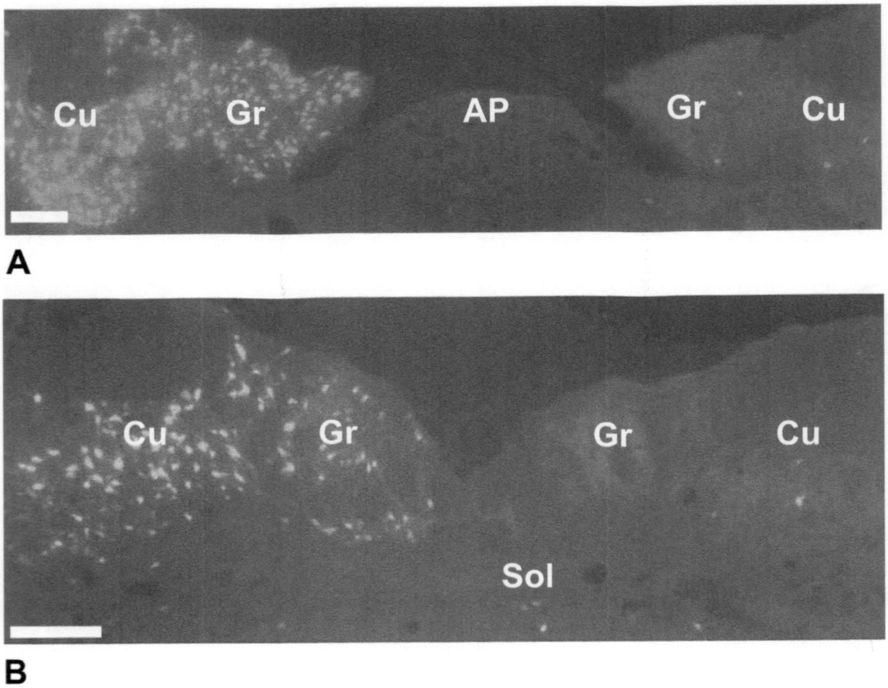

Fig. 10 A,B A large number of retrogradely labeled neurons are distributed throughout the contralateral gracile (*Gr*) and cuneate (*Cu*) nuclei (*left part of the pictures*), both rostrally **A** and caudally **B**. In the ipsilateral dorsal column nuclei, several labeled neurons are also seen. For orientation, the area postrema (*AP*) and the nucl. solitarius (*Sol*) are indicated. Scale bars: 250 μm

Fig. 11 (top) In the spinomedullary junction, a single retrogradely labeled neuron is seen in the most caudal contralateral gracile nucleus (*Gr*) and in the midline nucleus of Bischoff (*Bi*), respectively. In the spinal cord (*left half of the figure*) contralateral to the injection site distinctly retrogradely labeled neurons are seen in the lateral cervical nucleus (*LCN*) as well as in the lateral spinal nucleus (*LSN*). Within the grey matter, the retrogradely labeled neurons are scattered bilaterally. Note that in the ipsilateral cord neurons are located deep in the ventral horn (*arrowhead*). Also, two labeled cells are found within the lateral white matter (*arrow*). Scale bar: 200 μm

Fig. 12 (bottom) In the first cervical segment, the distribution of the retrogradely labeled neurons somewhat differs from that seen in the spinomedullary junction (Fig. 11). Here again, there are labeled neurons in the LCN and LSN contralateral to the injection site (*left half of the figure*). In lamina I, two labeled neurons are seen contralaterally and one ipsilaterally. In the deeper laminae distinctly retrogradely labeled neurons are found mainly in the medial grey matter, in a characteristic location of the STT cells. Bilaterally retrogradely labeled neurons are also found deep in the ventral horns. Scale bar: 200 μm

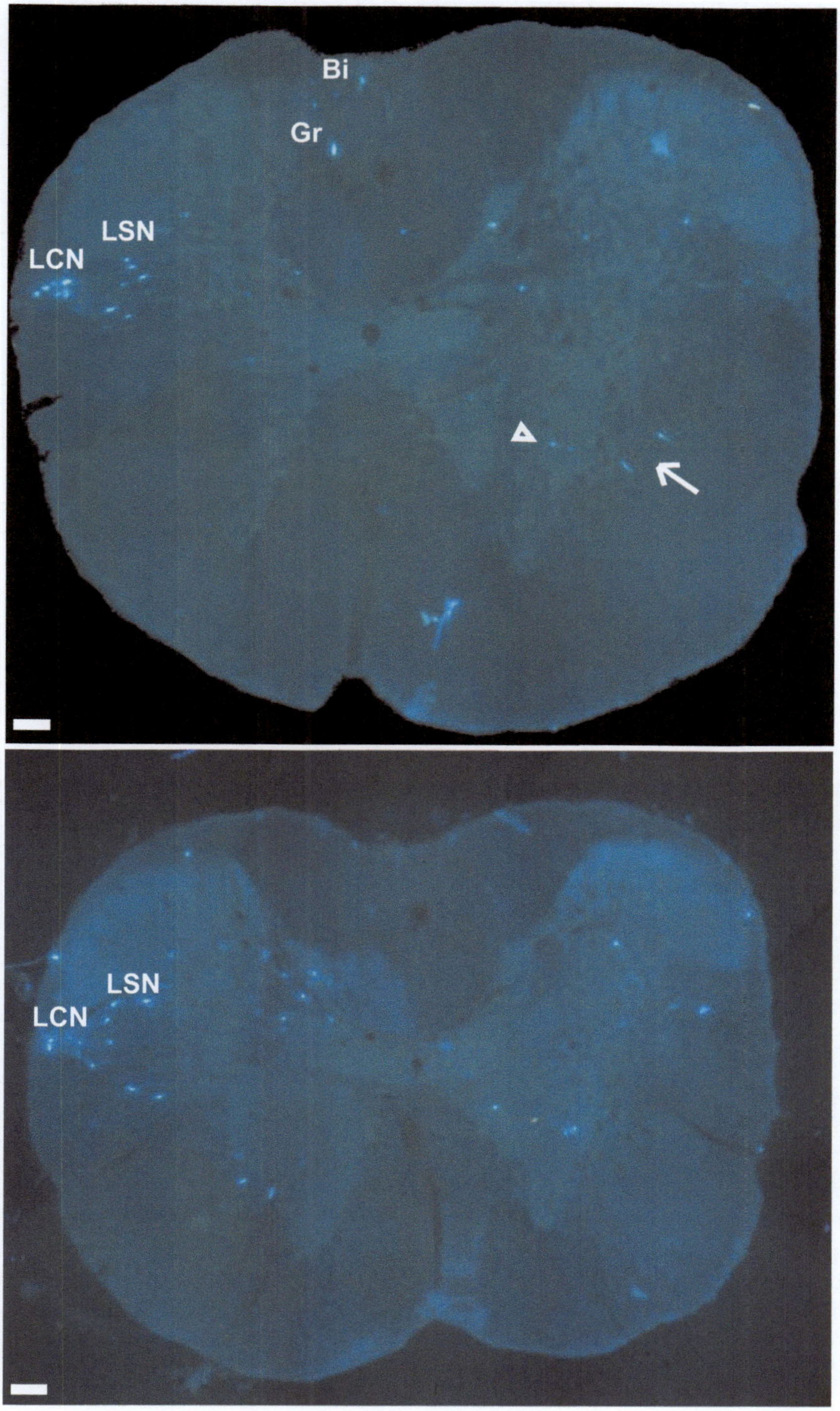
Bi
Gr
LSN
LCN
LSN
LCN

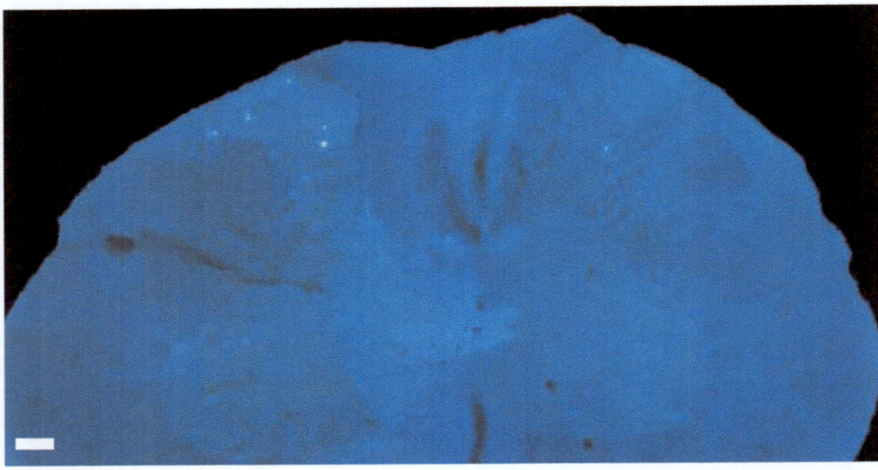

Fig. 13 Already at the level of the fifth cervical segment, the number of the retrogradely labeled neurons has drastically diminished. In the dorsal horn contralateral to the injection site (*left half of the figure*), labeled neurons are concentrated in the superficial laminae. Only one ipsilaterally projecting neuron is seen. Scale bar: 175 μm

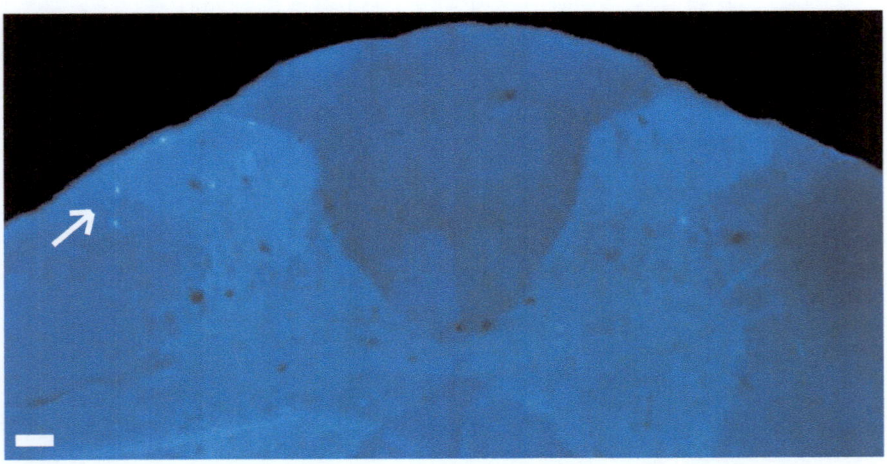

Fig. 14 In the first thoracic segment four retrogradely labeled neurons are depicted in the dorsal horn contralateral to the injection site (*left half of the figure*) and one is located in the LSN (*arrow*). There is one labeled neuron also seen in the ipsilateral DH. Scale bar: 125 μm

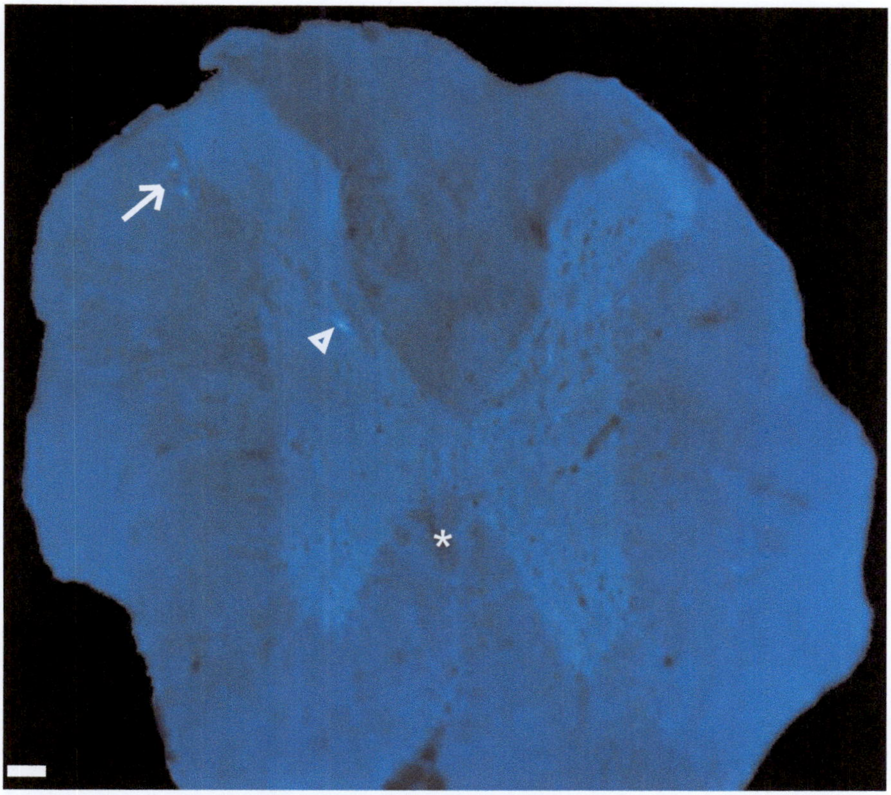

Fig. 15 At the level of fifth thoracic segment very few retrogradely labeled neurons are found. Two of them are located within the white matter contralateral to the injection site (*left half of the figure*) lateral of the DH in the LSN (*arrow*), and a large neuron is seen in the medial part of the deeper laminae (*arrowhead*). The central canal is indicated (*). Scale bar: 150 μm

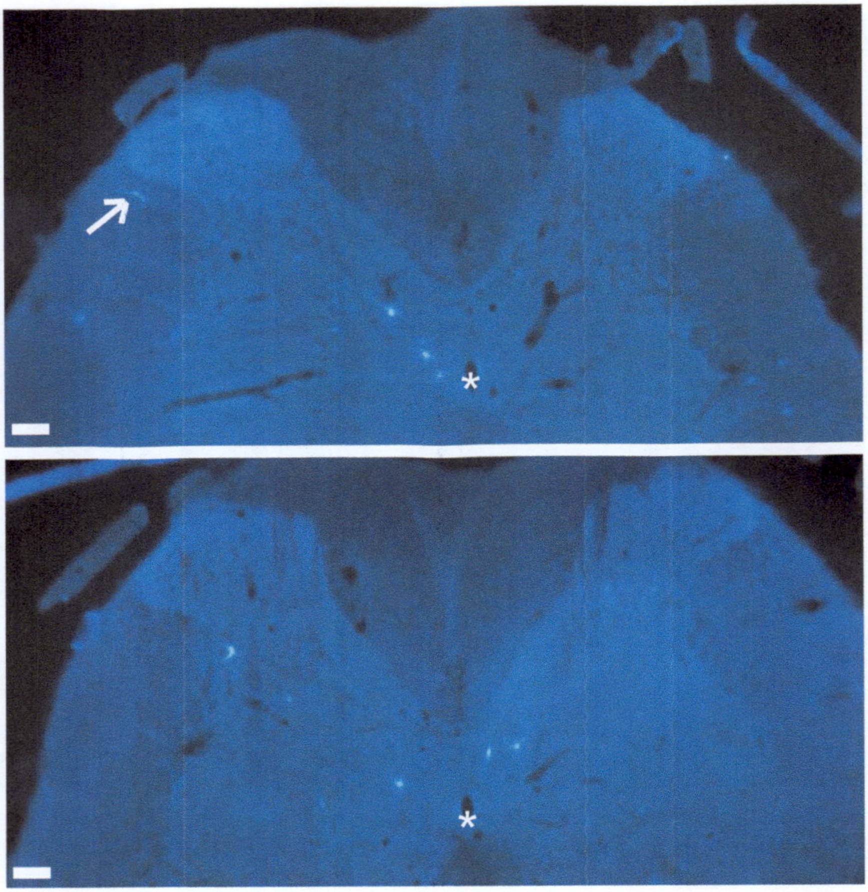

Fig. 16 A,B In the first lumbar segment the location of retrogradely labeled neurons differed between sections. **A** Contralateral to the injection site (*left half of the figure*), only labeled neurons are depicted at the base of DH, around the central canal (*) and a single one in the position of LSN (*arrow*). **B** Here, labeled neurons are seen bilaterally in the intermediate grey substance, one contralaterally in lamina III. Scale bars: 150 μm

Fig. 17 (top) In the fourth lumbar segment there are few retrogradely labeled neurons only contralateral to the injection site (*left half of the figure*). Two neurons are present in the base of DH near the central canal (*) and two small STT cells are seen in LSN (*arrow*). Scale bar: 150 μm

Fig. 18 A,B (bottom) In the first sacral segment, one strongly retrogradely labeled neuron is seen contralateral to the injection site (*left half of the figure*) in the deep DH. **B** In the enlargement of **A** the initial portion of the labeled axon is directed medially (*arrowheads*). **C, D** The first coccygeal segment is surrounded by the fascicles of the cauda equina (+). An elongated large neuron is seen in the medial portion of DH contralateral to the injection site (*left half of the figure*). **B** Enlargement of **A**. Scale bars: **A,C**, 250 μm; **B,D**, 500 μm

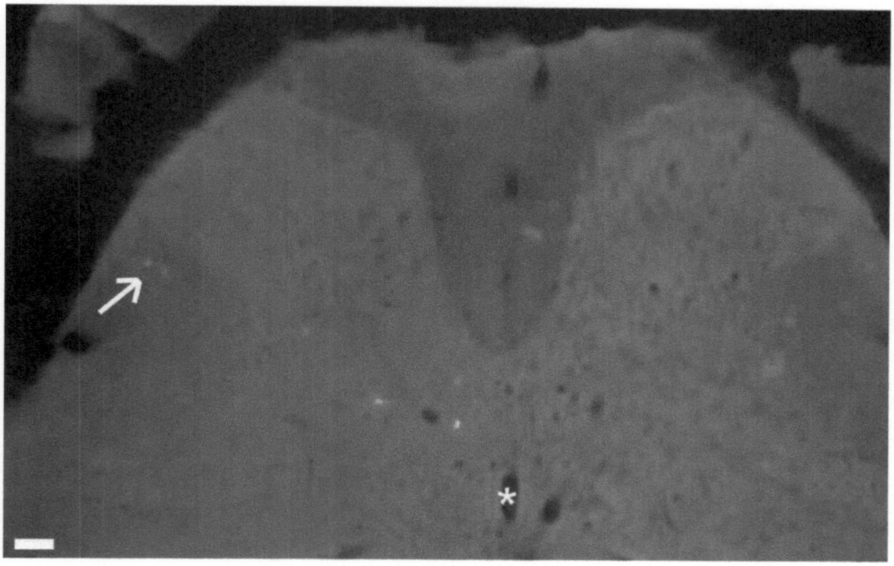

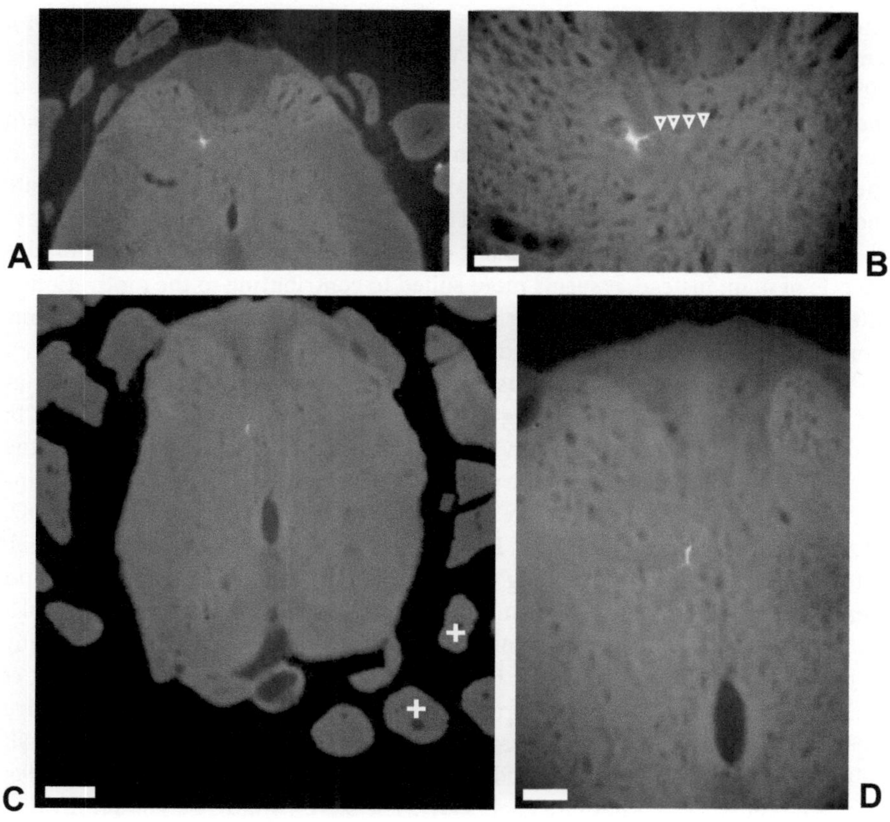
A
B
C
D

2.4.3 Pathways to Extrathalamic Structures

Several other pathways accompany the STT in the ventrolateral quadrant of the SC. These include the spinomesencephalic tract (SMT), the spinoparabrachial tract (SPbT), the spinoreticular tracts (SRT), and several more recently described spinolimbic tracts (Willis and Coggeshall 1991; Willis and Westlund 1997).

The SMT actually includes several projection systems that terminate in different mesencephalic areas. In primates, the neurons of origin are distributed similar to the STT neurons, e.g., in laminae I, IV–VI, and a few in the ventral horn and in lamina X (Willis et al. 1979; Mantyh 1982; Wiberg et al. 1987; Yezierski and Mendez 1991). The SMT neurons are glutamatergic (Yezierski et al. 1993; Azkue et al. 1998). Some SMT neurons emit collaterals to the lateral thalamus (Zhang D et al. 1990). The SMT projections terminate in the periaquaductal gray (PAG), nucl. of Darkschewitsch, nucl. interstitialis of Cajal, nucl. cuneiformis of the mesencephalic RF, nucl. intercollicularis, deep layers of the superior colliculus, area pretectalis, nucl. ruber, and probably in the cortical (polysensory, rather than acoustic) regions of the inferior colliculus (Bowsher 1957; Hassler 1960; Kerr 1975b; Mantyh 1982; Menetrey et al. 1982; Wiberg et al. 1987; Yezierski 1988; Blomqvist and Craig 1991; Yezierski and Mendez 1991; Bernard et al. 1995; Craig 1995). The SMT neurons project bilaterally (Wiberg et al. 1987; Blomqvist and Craig 1991; Craig 1995), and approximately 2% of the SMT cells in the rat project both ipsilaterally and contralaterally (Yezierski and Mendez 1991). According to Wiberg et al. (1987), the SMT projections from the caudal parts of the SC terminate in the caudal mesencephalon, and the projections from the cranial parts terminate more rostrally in the brainstem. The SMT is involved in nociception (Dougherty et al. 1999). However, it is not clear that it contributes to the sensory discriminative aspects of pain; instead, it seems more suited to contributing to the motivational, affective aspects of pain, as well to triggering activity in descending control systems (for details, see Willis and Westlund 1997).

There is growing evidence (at least in rodents) that the S(trigemino)PbT is a major nociceptive projection, rivaling in significance the STT (Bester et al. 1997b; Todd et al. 2000; Hunt and Mantyh 2001). This small region, surrounding the superior cerebellar peduncle at the pontomesencephalic transition, is densely innervated by ascending SC and STN axons (Hylden et al. 1985; Wiberg et al. 1987; Blomqvist et al. 1989; Bernard and Besson 1990; Craig 1992, 1995; Kitamura et al. 1993, 2001; Light et al. 1993; Slugg and Light 1994; Bernard et al. 1995; Feil and Herbert 1995; Allen et al. 1996; Yamashiro et al. 1998; Gauriau and Bernard 2002; Bourgeais et al. 2003). The cells of origin are located mainly in lamina I and many of them express the NK1 receptor (Ding et al. 1995b; Marshall et al. 1996; Yu et al. 1999; Todd et al. 2000; Bester et al. 2001), e.g., these cells receive a nociceptive input from SP-releasing PAs (Bernard et al. 1996; Craig 1996b; Hunt and Mantyh 2001, 2004; Todd et al. 2002). The SPbT is bilateral (Craig 1992; Bernard et al. 1996; Yamashiro et al. 1998; Kitamura et al. 2001). The S(trigemino)PbT is topically distributed (Bourgeais et al. 2003): parabrachial neurons excited chiefly by noxious stimulation of the face have their dendritic tree located primarily within the field of lamina I trigeminal projections, i.e., in the caudal portion of the parabrachial area, around the external medial and caudal part of the external lateral subnuclei; parabrachial neurons excited chiefly by noxious stimulation of the paw or the tail have their dendritic tree located primarily within the field of lamina I spinal projections, i.e., in parabrachial mid-extent, around the borderline between the external lateral and both the lateral crescent and the superior lateral subnuclei. The parabrachial nucleus projects heavily to the amygdala and the hypothalamus (Fulwiler and Saper 1984; Bernard and Besson 1990; Bester et al. 1997a; Gauriau and Bernard 2002). The spino-parabrachio-amygdalar/hypothalamic nocispecific multineuronal chain is probably concerned with the intensity of pain rather than its location or nature (Bernard et al. 1996; Bernard and Bandler 1998; Hunt and Mantyh 2001).

The involvement of the brainstem RF in pain conduction and modulation was studied intensively and reviewed by Hassler (1960), Bowsher (1976), Willis (1985), and Willis and Coggeshall (1991). The cells of origin of the SRT differ. Neurons in deeper laminae (V–VIII) project to the pontomedullary core: nucl. gigantocellularis, nuclei reticulares pontis oralis et caudalis (Kevetter and Willis 1982; Kevetter et al. 1982; Chaouch et al. 1983; Gauriau and Bernard 2002; for the delineation of the RF in the human brainstem see Paxinos et al. 1990; Koutcherov et al. 2004), and to nucl. reticularis lateralis (Menetrey et al. 1980, 1983). Lamina I neurons project to the dorsal central and ventrolateral reticular regions of the medulla oblongata (Craig 1995). Lamina I neurons project also to the catecholaminergic neurons of the brainstem, except for the dopaminergic group in the mesencephalon (substantia nigra and related nuclei: A8, A9, and A10 groups of Dahlström and Fuxe 1964). Craig (1992, 1995) and Westlund and Craig (1996) found that such axons project to noradrenergic and adrenergic groups in the ventrolateral medulla (A1 and C1), nucl. solitarius and the dorsomedial medullary RF (A2 and C2), the ventrolateral pons (A5), the locus coeruleus (A6), and the subcoerulear region (A7). Huber et al. (1999) encountered very few neurons in laminae II and III that project to the contralateral nucl. gigantocellularis. Nahin (1987) described several peptides in the SRT neurons. CCK-containing neurons were most common, while SP-containing cells were few. The data on the involvement of the RF in a nociceptive spino-reticulo-thalamic projection are contradictory. Bernard et al. (1990) think that the subnucl. reticularis dorsalis in the caudal medulla, which receives SRT axons and sends fibers to the parafascicular and ventromedial thalamic nuclei, could be involved in the control of pain processing. Especially Lima and Almeida (2002) argued that the subnucl. reticularis dorsalis is a prenociceptive center of the pain control system. Also, Villanueva et al. (1996, 1998) insist that the caudal RF is an important nociceptive relay to the thalamus, and the spino-reticulo-thalamic pathways may play an important role in distributing pain signals to the forebrain. On the other hand, Blomqvist and Berkley (1992) reexamined the spino-reticulo-diencephalic pathway in the cat, combining retrograde and anterograde tracing in order to study the extent to which SRT terminations and reticulodiencephalic neuronal perikarya overlap. They found SRT terminations mainly caudolaterally, while neurons projecting to the intralaminar thalamic nuclei and subthalamus were concentrated rostromedially. Thus, information conveyed from the SC to the RF appears to have access to the thalamus only by way of a few widely scattered neurons. According to Blomqvist and Berkley (1992), these results encourage less emphasis on a putative spino-reticulo-diencephalic pathway for pain. In the transmission of nociceptive spinal signals to the forebrain, a significant involvement of the pontomedullary noradrenergic neuronal groups could be ascribed, since they profusely innervate the thalamus, the hypothalamus, the amygdala, and the cerebral cortex (Aston-Jones et al. 1995; Westlund and Craig 1996). As Hassler proposed that the pallidum externum is reached by pain-conducting axons (see Fig. 31 in Hassler, 1960), recently Gauriau and Bernard (2004) established that the deep laminae in the rat SC project substantially to the globus pallidus and the substantia innominata. In addition to the multineuronal chains that convey nociceptive information to the hypothalamus and amygdala, there is growing evidence for the existence of direct spino(trigemino)hypothalamic and spino(trigemino)limbic tracts (Burstein and Giesler 1989; Burstein et al. 1990a, 1991, 1996; Cliffer et al. 1991; Katter et al. 1991, 1996; Iwata et al. 1992; Burstein and Potrebic 1993; Dado et al. 1994a, b, c; Zhang X et al. 1995c, 1999; Newman et al. 1996; Kostarczyk et al. 1997; Li et al. 1997; Yamashiro et al. 1998; Malick et al. 2000; Gauriau and Bernard 2004). The spinohypothalamic tract (SHT), at least in lower mammals, appears to be an unexpectedly massive projection. Burstein et al. (1990a) counted more than 9,000 retrogradely labeled neurons following selective injection of the tracer in the hypothalamus of rats. They found the greatest number of SHT neurons in the deep DH, followed by the LSN, superficial DH, and around the central canal; only a small number of spinohypothalamic neurons was found in the intermediate zone and in the ventral horn. Similar location of SHT and trigeminohypothalamic (THT) neurons, displaying SP receptor-immunoreactivity was reported by Li et al. (1997): most such neurons were located in lamina I. SHT in the cat has the same cells of origin as in the rat, but the projection appears to be smaller (Katter et al. 1991). SHT is present also in the monkey (Newman et al. 1996; Zhang X et al. 1999). All studies point out that the SHT is bilateral, predominantly crossed. The SHT axons terminate in most of the hypothalamic divisions: the lateral hypothalamus, posterior, dorsal, and periventricular areas, the dorsomedial, paraventricular, and suprachiasmatic nuclei, and the lateral and medial preoptic areas (Cliffer et al. 1991). In monkeys, the axons pass through the thalamus and then enter the hypothalamus (Zhang X et al. 1999). Similarly, in rats, the SHT axons run through the Po (Kostarczyk et al. 1997). The latter authors established that the SHT axons collateralize significantly in the brainstem, innervating numerous RF nuclei, nucl. ambiguus, nucl. solitarius, and Cu. Kostarczyk et al. (1997) conclude that through its widespread collateral projections, the SHT appears to be capable of providing nociceptive input to many areas that are involved in the production of multifaceted responses to noxious stimuli. Zhang X et al. (1995c) established that some SHT axons in the rat course through a long and complex path. After decussating in the hypothalamus, the axons descend in the ipsilateral Po, midbrain, pons, or even rostral medulla. Such axons may provide nociceptive information to a variety of nuclei throughout the diencephalon and brainstem bilaterally. Malick et al. (2000) found that most of the THT neurons are nociceptive. Their axons cross the midline and ascend until the level of supraoptic decussations in the lateral hypothalamus. More than a half of the axons recross the midline to reach the ipsilateral hypothalamus. The hypothalamic areas that receive trigeminal input are the lateral, perifornical, dorsomedial, suprachiasmatic, and supraoptic nuclei. The THT axons collateralize profusely: to the superior colliculus, substantia nigra, red nucleus, anterior pretectal nucleus, striatum, globus pallidus, and substantia innominata. According to Malick et al. (2000), the findings that non-nociceptive signals reach the hypothalamus through the direct THT route, whereas nociceptive signals reach the hypothalamus through both the direct and indirect routes, suggest that highly prioritized painful signals are transferred in parallel channels to ensure that this critical information reaches the hypothalamus, a brain area that regulates homeostasis and other humoral responses required for the survival of the organism.

Following the observation of Burstein and Giesler (1989) that the SC projects directly to the telencephalon, i.e., to the limbic structures such as nucl. accumbens and the septal nuclei, several papers confirmed and extended this unexpected finding. Cliffer et al. (1991) report a strikingly large number of structures that receive SC axons: ventral pallidum, globus pallidus, substantia innominata, basal nucleus of Meynert (cholinergic neuronal group that innervates profusely the cerebral cortex, the Ch4 group of Mesulam et al., 1984), amygdala, horizontal and vertical limbs of the diagonal band of Broca, medial and lateral septal nuclei, nucl. accumbens, and even the infralimbic and medial orbital cortex. The retrograde tracing experiments of Burstein and Potrebic (1993) indicated that the projection to the amygdala in the rat arises through the entire length of the SC. The number of spinoamygdaloid neurons is modest, and these cells are located bilaterally (mainly contralaterally) in the lateral reticulated area of the deep DH and around the central canal. These authors verified the projection to the orbital cortex but also pointed out that the number of spinocortical neurons is quite small. Newman et al. (1996) found spinal projections to the hypothalamus, ventral striatum, globus pallidus, amygdala, and the septal nuclei in rats and squirrel monkeys. They estimated that in both species the total number of terminals seen in the striatal and limbic areas was 50%–80% of the number seen within the thalamus. Following experimental tooth movement, Yamashiro et al. (1998) found in rats bilaterally Fos-expressing neurons in the periventricular hypothalamus and in the central nucleus of the amygdala. Presently, the laterocapsular part of the central amygdala is defined as the nociceptive amygdala because of its high content of nociceptive neurons (Bourgeais et al. 2001; Gauriau and Bernard 2002; Li and Neugebauer 2004).

2.5 Dorsal Column Nuclei and Nociception

Gr and Cu, their main afferent fibers traveling in the dorsal columns of the SC, and their efferent fibers traveling in the medial lemniscus, are a part of trisynaptic pathway traditionally thought to convey impulses concerned primarily with touch-pressure and kinesthesis (Foerster 1936; Willis and Coggeshall 1991; Snow and Wilson 1991; Parent 1996). It is appreciated that the PA neurons are pseudounipolar and their axons are myelinated. Giuffrida and Rustioni (1992) counted and measured thousands of retrogradely labeled SG neurons in rats that received a tracer in the DCN. They found that at every level, most labeled, i.e., projecting neurons are large.

Electrophysiological studies first addressed the role of the dorsal columns in mediating visceral pain (Amassian 1951; Rigamonti and Hancock 1978). More recently, Berkley et al. (1993) and Berkley and Hubscher (1995) have shown that the Gr neurons can be activated by distension of vagina, uterus, and colon, and half of the Gr cells that respond to cutaneous stimuli are also activated by uterine or vaginal distension. Apkarian et al. (1995) suggested that the DCN may be more important for visceral pain than is the STT. Willis and his colleagues published a series of papers that demonstrate the profound involvement of the DCN in the transmission of visceral pain (Al-Chaer et al. 1996a, b, 1997, 1998; Willis 1999; Nauta et al. 2000; Wang and Westlund 2001; Palecek et al. 2002, 2003a, b; Palecek and Willis 2003). The nociceptive inputs reach the DCN via two routes: (a) monosynaptic input from PA cells in the SG and (b) the pathway consisting of two neurons: a PA neuron and a neuron in the SC.

The classic monosynaptic nociceptive input was described repeatedly (Patterson et al. 1989, 1990; Garrett et al. 1992). According to Conti et al. (1990), the nociceptive input to the DCN may be mediated, though to a very limited extent, directly by way of small, substance P-containing PA neurons.

More important is the second route: via the so-called postsynaptic fibers traveling in the dorsal column. By this bisynaptic pathway, the central process of the PA neuron terminates upon a second-order projection neuron, located in the gray matter of the SC. The axons of these neurons—the postsynaptic fibers—reach the DCN (Rustioni 1973, 1974; Rustioni and Kaufman 1977; Cliffer and Giesler 1989; Cliffer and Willis 1994; Hirschberg et al. 1996; Wang et al. 1999). Rustioni (1977) and Rustioni et al. (1979) investigated the cells of origin of postsynaptic fibers in monkeys. The found that the fibers originate mainly from ipsilateral DH, particularly from its medial part at upper cervical levels and from a band of gray matter throughout the SC, largely corresponding to lamina IV and adjacent laminae. Large neurons along the lateral border of the ventral horn at lumbar levels may also contribute nonprimary afferents to the ipsilateral DCN. In the cat (Rustioni and Kaufman 1977), the cells of origin are numerous in the upper cervical, brachial, and lumbosacral SC, but are sparse in the thoracic segments. In the brachial and lumbosacral cord, the neurons of origin are mainly localized in lamina IV and more ventrally. According to Giesler et al. (1984), in the rat the postsynaptic dorsal column neurons constitute over 38% of the neurons that project to Cu, and approximately 30% that project to the Gr. In the lumbar segments, the cells of origin are located within a narrow band extending across the ipsilateral DH, subjacent to substantia gelatinosa. Hirschberg et al. (1996) reported a population of cells originating in lamina X and overlying dorsal commissural region at the sacral level of the rat SC. Similarly, Wang et al. (1999) found out that in the rat, neurons in the area adjacent to the central canal of the midthoracic or lumbosacral level of the SC send ascending projections to the dorsal, lateral rim of the Gr and the medial rim of Cu or the dorsomedial rim of the Gr, respectively. The non-PAs to the DCN ascend mainly in the dorsal columns and, to a lesser extent, in the dorsal part of the lateral funiculus both in monkeys (Rustioni et al. 1979) and in the rat (Giesler et al. 1984). The data on the role of the postsynaptic fibers in somatosensory processing are contradictory. Brown and Fyffe (1981) and Brown et al. (1983) indicated that this fiber system transmits cutaneous nociceptive and tactile information to the brain. On the other hand, Giesler and Cliffer (1985) remained skeptical that the postsynaptic fibers are involved in nociception. Also, according to Al-Chaer et al. (1996a, 1997) the dorsal columns play a minor role in relaying excitatory noxious cutaneous input to the VPL thalamic nucleus.

2.6 Cerebellum and Nociception

The cerebellum is regarded as a part of the CNS that is implicated mainly in motor behavior and its coordination. However, numerous studies showed a broad diversity of its functions (reviewed by Saab and Willis 2003). Data indicating that the cerebellum is also involved in nociception has been abundant in recent years, although Chambers and Sprague (1955a, b) described an analgesic effect following cerebellar cortical lesions. Siegel and Wepsic (1974) observed antinociceptive effects following electrical stimulation of the superior cerebellar peduncle in the monkey. Spiegel (1982) speculated that impulses generated by posterior column stimulation may lead to relief of pain and spasticity by activating the cerebellum.

The first reliable evidence that nociceptive stimulation evokes activity in pathways and neurons of the cerebellum was provided by Ekerot et al. (1987a, b). They reported that climbing fiber-evoked responses were recorded in Purkinje cells and as field potentials from the surface of the cerebellum upon stimulation of the ipsilateral superficial branch of the radial nerve. Similar data were reported by Wu and Chen (1990) following stimulation of C-fibers in the saphenous nerve. Ekerot et al. (1991) proposed that the cutaneous nociceptive input may be transmitted to the inferior olive by the postsynaptic dorsal column nuclei. McGonigle et al. (1996) found out that fibers containing substance P terminate upon neurokinin-1 receptor-immunoreactive neurons of the dorsal spinocerebellar tract that project to paravermal areas. Saab et al. (2001) examined the influence of cerebellar cortical stimulation on spinal nociceptive neurons that responded to noxious visceral and somatic stimuli. The stimulation increased the responses of all isolated cells to visceral stimuli (colorectal distension), while the effect on the responses to somatic stimuli was less clear. In addition, Saab and Willis (2001) found that Purkinje cells in the caudal vermis respond to nociceptive visceral stimulation in the form of early and delayed changes in activity, and proposed a negative feedback circuitry involving the cerebellum for the modulation of peripheral nociceptive events.

Recently, imaging studies on the nociceptive input to the cerebellum have also appeared. In positron emission tomography (PET) and functional magnetic resonance imaging (fMRI) studies, increases in blood volume or flow in the vermis and paravermal areas were reported during the perception of acute heat pain (Casey et al. 1994), deep cold pain (Casey et al. 1996), muscle pain (Svensson et al. 1997), and capsaicin-evoked pain and allodynia (Iadarola et al. 1998). Saab and Willis (2003) concluded that "... one central pillar (of pain research) is missing to confirm a role for the 'little brain' in pain: clinical data.... Whereas the 'little brain' may influence nociception, its grip on pain remains pliable."

2.7 Cortices Involved in Pain Perception and Thalamocortical Projections

There is a multiregional organization of supraspinal pain processing (Bromm and Lorenz 1998; Coghill et al. 1999; Treede et al. 1999; Hudson 2000; Peyron et al. 2000) and cortical areas involved in pain perception are the primary somatosensory cortex (SI), the secondary somatosensory cortex (SII), the insular (IC), the anterior cingulate (ACC), and the prefrontal (PC) cortices. The respective cortical areas differ functionally, as seen in electrophysiological and functional imaging studies: the sensory-discriminative aspect of pain (localization, intensity, duration, quality) is presented in SI and SII, receiving thalamic input from lateral thalamic nuclei, the motivational-affective aspect (subjective suffering, unpleasantness, aversive emotions), and the cognitive-evaluative aspects of pain are presented in the IC, ACC, and PC, receiving thalamic input from medial thalamic nuclei.

Primary Somatosensory Cortex The role of SI (located in the postcentral gyrus, Brodmann's areas 3, 1, 2) in pain perception has been a matter of dispute for decades. The early findings were largely negative. Head and Holmes (1911) reported that patients with long-standing cortical lesions did not show deficits in pain perception, which lead to an erroneous suggestion that the pain sensation takes place in the thalamus. During epilepsy surgery, Penfield and Boldrey (1937) performed electrical stimulation of patients' exposed SI and encountered only very few cases (11 out of more than 800 responses) that reported a sensation of pain. Single-cell recording in monkeys (Kenshalo et al. 1988) revealed only very few nociceptive neurons, and the authors concluded that their functional significance was uncertain. Also, the findings from human brain imaging studies have produced rather inconsistent results concerning the role of SI in pain perception (Bushnell et al. 1999; Craig 2003a). Despite certain controversies, an increasing number of PET and fMRI studies found an activation of SI during painful stimuli (Casey et al. 1994; Coghill et al. 1994; Andersson et al. 1997; Torebjörk 1997; Derbyshire and Jones 1998; Porro et al. 1998; Davis 2000; Wiech et al. 2001), also corroborating electrophysiological findings (Drushky et al. 2000; Kanda et al. 2000). According to Craig (2003a, d), nociceptive activation near the central sulcus in humans probably occurs in area 3a (where the thalamic VMpo projects), but its location is below the level of PET resolution. Bushnell et al. (1999) suggest that in SI primarily the sensory-discriminative aspect of pain is presented. Two classes of neurons are activated in SI: neurons with a wide dynamic range react already to stimuli that are not painful; however, they show the highest activity to painful stimuli (Chudler et al. 1990). They have large receptive fields and probably code pain intensity. Specific nociceptive neurons only react to painful stimuli. They have small receptive fields, are somatotopically located in the postcentral gyrus and enable the determination of the localization, intensity, and temporal attributes of the painful stimuli. The SI neurons get their afferents from the lateral thalamic nuclei (VPL, VPM, VPI; in primates and humans also from VMpo; Willis 1997), and also heavily project back to these nuclei. The thalamocortical projections are excitatory glutamatergic (Kharazia and Weinberg 1994). Lesions of the respective thalamic nuclei, the thalamocortical connections or of SI result (besides loss of somatosensory function) in a dramatic decrease in temperature and pain perception (Bassetti et al. 1993; Leijon et al. 1989). But there is no complete analgesia. Nevertheless, pain is still interpreted as uncomfortable and unpleasant (Ploner et al. 1999).

Secondary Somatosensory Cortex SII is located just lateral and slightly anterior to the lateral end of the central fissure in the human brain, roughly occupying Brodmann's area 43 and parts of area 40. In contrast to SI, SII neurons do not seem to be involved in discrimination of location and/or intensity of painful stimuli, but seem to have an important role in recognition, learning, and memory of painful events (Schnitzler and Ploner 2000). A number of studies found significant pain-related activation of SII with functional imaging and electrophysiological methods (Talbot et al. 1991; Casey et al. 1994; Coghill et al. 1994; Oshiro et al. 1998; Xu et al. 1997; Davis 2000; Druschky et al. 2000; Kanda et al. 2000; Treede et al. 2000), mostly bilaterally. The SII neurons get their mostly bilateral afferences from the lateral thalamic nuclei partly different from those projecting to SI, namely from the VPI and the dorsal part of the Po, thus indicating an anatomical and functional segregation of the SI- and the SII-nociceptive pathways. Additionally, SII is reciprocally connected to SI. Nevertheless, the function of SII in pain processing is still unclear. Lenz et al. (1997) proposed that SII may play a key role in relaying nociceptive information to the IC and the temporal lobe limbic structures, providing fast access to pain-related learning and memory.

Insular Cortex Functional imaging studies showed increased blood flow of the insular cortex during painful stimuli, either contralaterally or bilaterally (Casey et al. 1994; Coghill et al. 1994; Andersson et al. 1997; Derbyshire and Jones 1998; Treede et al. 1999, 2000; Davis 2000; Sawamoto et al. 2000). It is not yet clear whether the anterior (Brodmann's area 13) or posterior insular cortices (Brodmann's areas 14–16) are mainly involved in pain perception (Craig 2003c, d). Moreover, patients with lesions of the IC had an elevated pain tolerance and loss of or inadequate emotional reactions to painful stimuli although recognizing pain (asymbolia for pain; Bertier et al. 1988, Greenspan et al. 1999). The IC gets thalamic afferents from the VMpo, the mediodorsal (MD), and intralaminar thalamic nuclei (Craig et al. 1994; Craig 1996a) and from SII, and projects to limbic structures such as the amygdala and the perirhinal cortex. Also, these connections speak in favor of the importance of the IC in the motivational-affective aspect of pain and in autonomic reactions to noxious stimuli.

Anterior Cingulate Cortex The cingulated cortex is involved in cognition and emotion. Both functions are located in different anatomical subareas. The subarea involved in the motivational-affective aspect of pain is most probably located in the rostral part of Brodmann's area 24 and the adjoining area 32. Patients with lesions of the ACC lost the emotional reactions to painful stimuli although pain could be further correctly localized. In the ACC, pain-receptive neurons were found with large, often bilateral receptive fields not allowing localizing information. Significantly increased functional activity of the ACC was robustly found in many imaging studies (Casey et al. 1994, 1996; Coghill et al. 1994; Derbyshire and Jones 1998; Bromm et al. 2000; Casey 2000; Davis 2000; Hudson 2000) mostly described in the hemisphere contralateral to the painful stimulus. The ACC gets thalamic afferents from the VMpo, the MD, and intralaminar nuclei, from the IC and PC, and projects to the amygdala, the mediodorsal thalamic nuclei, the PAG, motor nuclei of the brainstem, and the IC thus being involved in motivational-affective aspects of pain and in conditioned fear reaction. As the ACC can modulate the affective aspect of sensory perception by pain expectation it is also involved in mediating the affective components associated with attention and anticipation of upcoming noxious stimulation (Sawamoto et al. 2000). In this respect, it is interesting to note that hypnotic suggestion can selectively alter the unpleasantness of noxious stimuli in parallel with reduced pain-evoked activity within the ACC (Rainville et al. 1997). Thus the ACC may have a pivotal role in interrelating attentional functions with that of establishing emotional valence and response properties (Price 2000).

Prefrontal Cortex There are still some doubts with respect to the function of the PC in pain perception. The PC is rather believed to function as a supervisory attention system (Andersson et al. 1997) and to be correlated with the cognitive-evaluative aspect of pain. Functional imaging studies, however, described activation of parts of the PC (probably Brodmann's areas 9 and 10) during the painful stimuli. Interestingly, mostly the right hemisphere showed increased activity irrespective of the side of stimulation (Derbyshire and Jones 1998). Patients with unilateral lesions of the PC show changes in both the sensory-discriminative and the motivational-affective aspects of pain. The PC gets thalamic afferents from the VMpo, the MD and intralaminar nuclei, and projects to the MD and the ACC.

Sewards and Sewards (2002) proposed that separate sensory and hedonic representations exist in each of the primary structures of the somatosensory system, including brain stem, thalamic, and cortical components. They think that in rodent primary somatosensory cortex, a hedonic representation can be found in laminae Vb and VI. In carnivore and primate primary and secondary somatosensory cortical areas no hedonic representation exists, and the activities of neurons in both areas represent the sensory aspect exclusively. However, there is a hedonic representation in the posterior part of the insular cortex, bordering on the retroinsular cortex, that receives projections from the thalamic areas in which hedonics are represented. According to Sewards and Sewards (2002), these segregated components are related to the subjective awareness of pain.

Motor Cortex Interestingly, motor cortex stimulation has been shown to be benevolent for chronic pain suppression. Nearly 300 cases of motor cortex stimulation have been published. Although the results were variable, it was applied successfully in central post-stroke pain and in trigeminal neuralgia. The electrode is placed epidurally over the precentral gyrus. By stimulating the precentral cortex, increased neuronal activity was found in the ventroanterior and ventrolateral nuclei of the thalamus. Computer modeling can predict the immediate bioelectrical effects of the motor cortex stimulation (see Manola et al. 2005 for overview and modeling).

2.8 Descending Modulatory Pathways

The communication of Reynolds (1969) that he was able to perform abdominal surgery in rats without chemical anesthesia, but instead stimulation of the midbrain PAG, was followed by a veritable boom of investigations on the descending analgesia systems. The considerable body of literature was reviewed by Basbaum and Fields (1984), Willis (1984), Besson and Chaouch (1987), Willis and Coggeshall (1991), Light (1992), Wang and Nakai (1994), Beitz (1995), Stamford (1995), Willis et al. (1995), Willis and Westlund (1997), Fields and Basbaum (1999), Fields (2000), Lima and Almeida (2002), and Suzuki et al. (2002). Therefore, only a concise review will be presented here.

The efferent connections of the PAG to the SC are indirect. The PAG neurons project to the serotoninergic raphe nuclei of the medulla oblongata and to the noradrenergic nuclei in the dorsolateral pons (Van Bockstaele et al. 1991; Bajic and Proudfit 1999). Both the catecholaminergic and indolaminergic neuronal groups project heavily to the SC and to the STN.

From the serotoninergic groups, the largest contribution of raphespinal connections is provided by nucl. raphe magnus, followed by the pallidus, obscurus and pontis raphe nuclei (Bowker et al. 1981, 1983; Steinbusch 1981; Willis 1984; Kwiat and Basbaum 1990; Jones and Light 1990, 1992; Jones et al. 1991; for the topography of the raphe nuclei in the human brainstem, see Törk and Hornung 1990). The serotoninergic nucl. raphe dorsalis, located in the midbrain, also participates in antinociception, however not with a direct raphespinal connection; it is rather involved both in ascending and descending pain inhibitory systems (Wang and Nakai 1994). Polgar et al. (2002) showed that the serotonin-containing axons in the SC selectively innervate the lamina I projection neurons that possess the NK1 receptor.

The noradrenergic connections to the SC arise in the locus coeruleus, subcoeruleus nucleus, and nucleus of Kölliker-Fuse (Westlund and Coulter 1980; Holstege and Kuypers 1982; Stevens et al. 1982; Westlund et al. 1983, 1984; Kwiat and Basbaum 1990; Clark and Proudfit 1991; Yeomans and Prodfit 1992; West et al. 1993; Zhang C et al. 1997; Tsuruoka et al. 2003). The projections are bilateral, predominantly crossed, and mainly laminae I, II, and V are innervated. Zhang C et al. (1997) stated that there is a predominantly inhibitory role on nociceptive transmission at the SC level by descending noradrenergic fibers, and a facilitatory role on the responsiveness of the thalamic parafascicular nucleus to noxious inputs by ascending locus coeruleus axons. Tsuruoka et al. (2003) found out that a unilateral inflammation of the hind paw in rats results in bilateral activation of locus coeruleus, followed by descending modulation.

The neurochemistry of the transmitters and receptors in the multineuronal antinociceptive pathway arising in the PAG is very complex (Bowker et al. 1983; Cui et al. 1999). Along with serotonin and noradrenaline, also endogenous opiates and the amino acids glutamate, GABA, and glycine are clearly involved (Willis 1985; Willis and Coggeshall 1991; Stamford 1995; Willis and Westlund 1997; Lima and Almeida 2002).

The pretectal area is regarded as a part of the visual system. However, the connections of the anterior pretectal nucleus suggest that it is a part of the somatosensory system (Berkley et al. 1986; Wiberg et al. 1987; Foster et al. 1989; Yoshida et al. 1992; Terenzi et al. 1995). Stimulation in the anterior pretectal nucleus results in long-lasting antinociception without aversive side effects (Rees and Roberts 1993). Again, the antinociceptive impulses, arising in the anterior pretectal nucleus, are mediated via descending multineuronal chains, involving the deep mesencephalic nucleus, the pedunculopontine tegmental nucleus (the cholinergic Ch5 group of Mesulam et al. 1984, 1989), and the noradrenergic and serotoninergic neurons in the pons and medulla (Terenzi et al. 1991, 1992, 1995; Wang et al. 1992; Zagon et al. 1995).

In human patients, stimulation of the VPM and VPL thalamic nuclei is followed by a reduction in pain in postherpetic neuralgia (PHN), thalamic syndrome, and facial anesthesia dolorosa (Turnbull et al. 1980). Gerhart et al. (1983) found that stimulation in the VPL causes an inhibition of primate STT neurons. Such inhibition might result from antidromic activation of STT axons that emit collaterals to nucl. raphe magnus and to the PAG. Also, the stimulation of the SI region of the monkey cerebral cortex causes the inhibition of STT neurons (Yezierski et al. 1983). However, the cortical inhibition acts mainly on the responses to innocuous mechanical stimulation, rather reducing nociceptive responses (Yezierski et al. 1983; Zhang D et al. 1991).

Although the focus of investigation has been on the inhibitory modulation of spinal nociceptive processes, data are accumulating that brain stem stimulation can also enhance spinal nociceptive processes (Porreca et al. 2002). Fields (1992) suggested that descending facilitatory influences could contribute to chronic pain states. Later, Urban and Gebhart (1999) stated that such influences were important to the development and maintenance of hyperalgesia. Several studies indicate that the rostroventromedial medulla is a crucial relay in the persistence of descending facilitation of noxious stimuli (Porreca et al. 2002).

The spinal neurons that express the NK1 receptor appear to play a pivotal role in regulating descending systems that modulate activity of nociceptive dorsal horn neurons (Mantyh and Hunt 2004; Khasabov et al. 2005).

3 Neuropathic Pain

While the acute nociceptive pain is a necessary defense mechanism that warns against damage to the organism, chronic pain can be so deleterious that patients not rarely prefer death. The nociceptive ("good") pain is essential for survival but the chronic ("bad") pain serves no defensive, helpful function. There is no biological advantage but only suffering and distress. Acute pain is produced by the physiological functioning of the normal nervous system. The chronic, maladaptive pain typically results from damage to the nervous system (peripheral nerve, PA neuron, CNS) and is known as neuropathic pain (Basbaum 1999; Dworkin and Johnson 1999; Woolf and Salter 2000; Bridges et al. 2001; Hunt and Mantyh 2001; Zimmermann 2001; Scholz and Woolf 2002; Woolf 2004; Tsuda et al. 2005).

The spectrum of NP covers a variety of disease states and presents in the clinic with a variety of symptoms (Woolf and Mannion 1999; Bridges et al. 2001). Several etiologies of peripheral nerve injury might result in NP: PHN (Dworkin et al. 1997; Dworkin and Johnson 1999), traumatic injury (Schwartzman and Maleki 1999; Rodriguez-Filho et al. 2003), phantom limb pain (Nicholajsen and Jensen 2001), diabetes (Boulton and Ward 1986; Calcutt 2002; Khan et al. 2002; Simmons and Feldman 2002; Kapur 2003; Spruce et al. 2003), and malignancy (Schwei et al. 1999; Regan and Peng 2000; Cain et al. 2001; Farrar and Portenoy 2001; Clohisy and Mantyh 2003; Sabino et al. 2003). Despite its varied etiologies, NP conditions share certain clinical characteristics: spontaneous, continuous pain, usually of a burning character; paroxysmal (shooting, lancinating) pain; evoked pain to various mechanical or thermal stimuli such as allodynia and hyperalgesia. Hyperalgesia is an increased pain response to a suprathreshold noxious stimulus and is a result of abnormal processing of nociceptor input. Allodynia is the sensation of pain elicited by a non-noxious stimulus and can be produced in two ways: by the action of low threshold myelinated Aβ-fibers on an altered CNS, and by a reduction in the threshold of nociceptive fibers in the periphery. The fact that pain is often located in hypoesthetic or anesthetic areas may appear paradoxical and implies that NP not only depends on the genesis of nociceptive messages from nociceptors, but may depend on other mechanisms as well, in contrast to nociceptive pain (Attal and Bouhassira 1999).

That terminals of uninjured PA neurons terminating in the DH can collaterally sprout was first suggested by Liu and Chambers (1958), but was disputed by numerous investigators (Mannion et al. 1996; Wilson and Kitchener 1996). Woolf and colleagues presented a series of reports on the topographic reorganization of the SC PAs following chronic NP (Fitzgerald et al. 1990; Woolf et al. 1992, 1995; Coggeshall et al. 1997, 2001; Doubell et al. 1997, 1999; Mannion et al. 1998; Mannion and Woolf 2000; Tandrup et al. 2000; Woolf and Salter 2000; Decosterd et al. 2002; Sabino et al. 2003). Peripheral nerve injury results in a rearrangement of the highly ordered laminar termination of PAs within somatotopically appropriate regions of the DH. As described above, large myelinated mechanoceptive Aβ-axons normally terminate in laminae III–VI, thin myelinated nociceptive Aδ-fibers in laminae I and V, and the thinnest, unmyelinated C-axons in lamina II. Peripheral axotomy causes long-lasting sprouting of A-fibers into lamina II, an area in which they do not normally terminate. Intracellular injections of tracers show that at least some of these fibers are Aβ-afferents from lamina III. This A-fiber sprouting into lamina II appears to be a result of at least two phenomena. The first is the presence of vacant synaptic sites within the superficial laminae following the transganglionic degeneration of C-axons; the second is the induction of a regenerative capacity in the injured neurons (Mannion et al. 1996). Intrathecally supplied neurotrophic factors, which may act as C-fiber "therapy," can prevent A-fiber sprouting (Bennett 1994). The functional importance of A-fiber sprouting is that lamina II begins to receive information about non-noxious stimuli. This information may be misinterpreted by the CNS as noxious: an anatomical substrate for mechanical allodynia (Woolf and Doubell 1994; Attal and Bouhassira 1999; Woolf and Mannion 1999; Bester et al. 2000). The findings of Woolf and colleagues concerning sprouting of A-axons in the superficial laminae were confirmed by others (Koerber et al. 1999; Nakamura and Myers 1999; Kohama et al. 2000), and several reports on regenerative sprouting following nerve injury also appeared (McMahon and Kett-White 1991; Cameron et al. 1992; Florence et al. 1993; LaMotte and Kapadia 1993; Florence and Kaas 1995; Darian-Smith and Brown 2000; Darian-Smith 2004). On the other hand, Tong et al. (1999) demonstrated that in monkey and rat, a subpopulation of mainly small PA neurons acquires the capacity to take up certain tracers (cholera toxin) after axotomy, a capacity normally not associated with these SG neurons. Thus, after peripheral axotomy, cholera toxin is a marker not only for large but also for small (nociceptive) neurons, thus possibly also for both myelinated and unmyelinated PAs. According to Blomqvist and Craig (2000), such phenotypic changes mean that axonal sprouting may be less pronounced than originally assumed. It is clear that both peripheral and central pathophysiological mechanisms contribute to PHN pain. Some PHN patients have abnormal sensitization of unmyelinated cutaneous nociceptors (irritable nociceptors) and minimal sensory loss. Other patients have pain associated with small fiber deafferentation. In such patients, pain and temperature sensations are profoundly impaired but mechanical stimuli can produce severe pain (allodynia). In these patients, allodynia may be due to the formation of new connections between non-nociceptive, thick (Aβ) PAs and central pain transmission neurons. The third class of patients complain of severe spontaneous pain without hyperalgesia or allodynia, and according to Fields et al. (1998), such patients presumably have lost both large- and small-diameter fibers, and the pain is likely due to increased spontaneous activity in deafferented central neurons and/or reorganization of central connections. The central sensitization is an activity-dependent functional plasticity that results from activation of different intracellular kinase cascades leading to the phosphorylation of key membrane receptors and channels, increasing synaptic efficacy (Woolf and Mannion 1999; Ji and Woolf 2001).

The experiments with animal models of NP (Bennett and Xie 1988; Seltzer et al. 1990; Kim and Chung 1992; Chacur et al. 2001; Decosterd and Woolf 2000; Khan et al. 2002; Mantyh et al. 2002; Rodriguez-Filho et al. 2003) have considerably increased our knowledge on the neuroanatomical and neurochemical plasticity in the CNS. Sugimoto et al. (1989) reported that following a ligature around the ischiadic nerve in rats, there was a bilateral increase in the number of neurons in the lumbar region of the SC showing signs of degeneration (pyknosis and hyperchromatosis); however, the ipsilateral increase was significantly greater. They also noted that daily doses of strychnine in neuropathic animals significantly increased the incidence of degenerative neurons, suggesting that excessive excitation, which can be exacerbated by strychnine-induced disinhibition, is one mechanism underlying the appearance of such cells. Further, Sugimoto et al. (1990) observed massive transsynaptic degeneration of interneurons in lamina II, and suggested that these neurons die through an excitotoxic mechanism.

The sympathetic nervous system probably plays a role in a relatively low subset of patients with PHN (Nurmicco et al. 1991; Sato and Perl 1991; Jänig 1996; Attal and Bouhassira 1999; Wu et al. 2000; Kress and Fickenscher 2001). The normal PA pseudounipolar neurons (except for the mesencephalic nucleus of the trigeminal nerve) receive no synaptic input (Zenker and Neuhuber 1990; Willis and Coggeshall 1991). However, following peripheral nerve lesion a sprouting of sympathetic noradrenergic fibers takes place (McLachlan et al. 1993; Chung et al. 1996, 1997; McLachlan and Hu 1998; Ramer et al. 1999). The sympathetic fibers, which normally innervate the blood vessels in the ganglia, now form basket-like structures around PA somata without establishing synaptic contacts with them, as is usual in normal tissues (autonomic nonsynaptic terminals). Zhou et al. (1999) suggest that satellite cell-derived nerve growth factor and neurotrophin-3 are involved in the induction of the sympathetic sprouting. However, the sympathetic sprouts predominantly form pericellular baskets around the large SG neurons that do not transmit pain information (Ramer et al. 1998; Zhou et al. 1999), and the density of sympathetic sprouts in the SG does not correlate with NP intensity (Baron et al. 1999). Sympathectomy has been shown to alleviate allodynia in animal models, some patients with causalgia respond positively to sympathicolytic procedures, and injection of epinephrine in a stump neuroma may induce intense pain (Chabal et al. 1992; Choi and Rowbotham 1997; Attal and Bouhassira 1999). Similarly, intracutaneous applications of adrenaline and phenylephrine have been shown to increase spontaneous pain and allodynia in the affected area of PHN patients (Choi and Rowbotham 1997). However, most of the patients do not demonstrate significant benefit from various sympathicolytic procedures (Nurmikko et al. 1991; Kingery 1997; Attal and Bouhassira 1999; Ochoa 1999; Ochoa and Verdugo 2001; Mailis and Furlan 2003), suggesting that the role of the sympathetic nervous system may have been overemphasized. Moreover, Mailis and Furlan (2003) point out that the complications of the sympathicolytic procedure may be significant, in terms of both worsening the pain or producing a new pain syndrome.

Along with neuroanatomical plasticity, the NP is accompanied with sometimes profound neurochemical plasticity, especially in the PAs. Experimental studies (Abbadie et al. 1996; Basbaum 1999; Schwei et al. 1999; Hunt and Mantyh 2001; Gardell et al. 2003; Hains et al. 2003a, b) strongly suggested that often there are distinct differences in the neurochemical changes that occur in the PA neurons and in the SC in neuropathic, inflammatory, and cancer pain states. In NP models (injuring the spinal nerve by means of cutting, crushing, or ligating) in the PA neurons there is a down-regulation of CGRP, SP, isolectin B4, and fluoride-resistant acid phosphatase, combined with up-regulation of glutamate, galanin, NPY, VIP, dynorphin, and GAP-43 (Jessel et al. 1979; Bennett et al. 1989; Noguchi et al. 1990; Cameron et al. 1991; Villar et al. 1991; Donnerer and Stein 1992; Al-Ghoul et al. 1993; Zhang X et al. 1993a, 1995a, b; Hökfelt et al. 1994; Ma and Bisby 1998; Miki et al. 1998; Honore et al. 2000b; Blakeman et al. 2003). On the other hand, in a model of persistent inflammatory pain, Honore et al. (2000b) encountered increases in SP and CGRP, and Segond von Banchet et al. (2002) showed increased up-regulation of neurokinin 1 and bradykinin 2 receptors in DRG neurons subsequent to antigen-induced arthritis.

In animal models of NP, a significant up-regulation of NOS was found in the PA neurons (Verge et al. 1992; Zhang X et al. 1993b; Steel et al. 1994; Choi et al. 1996; Shi et al. 1998; Luo et al. 1999). In addition, Gordh et al. (1998) encountered NOS up-regulation also in the SC gray matter, ipsilateral to the ligated spinal nerve. In all probability, also carbon monoxide plays a role in nociceptive processes, since Gordh et al. (2000) found an up-regulation of its synthesizing enzyme, the heme oxygenase. Zhang X et al. (1998) suggested that one factor underlying the insensitivity of NP to opioid analgesics could be due to a marked reduction in the number of mu-opioid receptors both in the axotomized primary sensory neurons and in the lamina II interneurons. Furthermore, after sciatic nerve ligation in the mouse Narita's group (Narita et al. 2000, 2004) demonstrated an up-regulation of various protein kinase C isoforms in the superficial layers of the DH, hypothesizing that these molecules are implicated in the sensitization of synaptic transmission associated with persistent pain.

Actual studies in neuropathic rats using the chronic constriction injury (CCI) model of the sciatic nerve reveal that important changes also take place in the respective muscles (Gradl et al. 2005). Muscles with impaired innervation react with apoptosis of their fibers. However, at present it is unclear how apoptosis of the muscle tissue contributes to neuropathic pain.

Furthermore, in the CCI model the invasion of T lymphocytes into the injured nerve was found to be correlated with neuropathic pain, whereas athymic nude rats, which lack mature lymphocytes, develop a significantly reduced allodynia and thermal hyperalgesia compared to normal rats (Moalem et al. 2004). Transfer of cytokine-producing T lymphocytes from CCI rats into nude rats enhanced pain hypersensitivity in the recipients, speaking in favor of the T cell immune response as a potential and important target for the treatment of NP (Moalem et al. 2004).

3.1 Central Changes Consequent to Peripheral Nerve Injury

Peripheral nerve injury in humans may result in clinical pain, including enhanced responsiveness to noxious stimuli (hyperalgesia) and the sensation of pain in response to innocuous stimuli (allodynia, Willis 1992). The two phenomena may involve different mechanisms, but an injury-triggered discharge in small-caliber PA fibers leading to hypersensitivity of DH neurons may occur at least at initial stages of both (McMahon et al. 1993; Thompson et al. 1993; Woolf and Doubell 1994). This increased excitability can be blocked by glutamate antagonists (Woolf and Thompson 1991; Liu and Sandkühler 1995), supporting release of glutamate by these fibers and a primary role for glutamatergic transmission in hypersensitivity (Willis 2001, 2002). Thus, better understanding of the mechanisms of hypersensitivity may be gained by studying the effects of peripheral injury on glutamate and its receptors. Glutamate receptors in the SC are down-regulated bilaterally following unilateral inflammation of the paw in rats, possibly as a result of indirect effects of the lesion (Pellegrini-Giampietro et al. 1994; Kus et al. 1995). On the other hand, immunocytochemical evidence suggests ipsilateral up-regulation of AMPA receptors in superficial laminae of the DH following chronic nerve ligation (Harris et al. 1996). While these apparent discrepancies may be explained on the basis of differences in the experimental models, none of these studies provides direct evidence that changes in glutamate receptors occur at synapses of PA terminals. Recent advances in postembedding immunocytochemistry made it possible to address this question at the first brain synapse (Phend et al. 1995; Kharazia et al. 1996; Matsubara et al. 1996; Popratiloff et al. 1996a). The present study is focused on the lamina II, because this is the region where the basic mechanisms responsible for the processing of nociceptive stimuli reside and where peripheral fibers involved in central sensitization after injury terminate (Woolf and Doubell 1994). Section of a peripheral nerve was chosen as the experimental model, because this procedure is known to result in hyperexcitability of DH cells, perhaps triggered by ectopic discharge at the neuroma or in SG (Devor 1994), and because it is highly reproducible from animal to animal.

Changes in AMPA Receptor Expression in Substantia Gelatinosa After Sciatic Nerve Lesion Sections reacted for FRAP exhibited a dense band of reaction product in the superficial dorsal horn on the control side (Popratiloff et al. 1998a). A portion of this band, corresponding to the representation of the sciatic nerve, was attenuated or absent on the lesioned side. In contrast to this prominent effect, only modest changes were seen in immunoreactivity for GluR2/3 on the two sides using conventional confocal microscopy. These included weakly increased staining intensity for somata, dendrites and poorly defined neuropil on the lesioned side. The mean intensity of immunofluorescence over lamina II on the lesioned side (as measured in ten 25-μm-thick sections from two rats) was only 7% greater than that on the control side. However, more detailed image analysis revealed significant changes in staining, especially a substantial increase in the number of very bright pixels on the lesioned side (Popratiloff et al. 1998a). Though consistent with an up-regulation of glutamate receptor protein, it was not possible from LM data to establish whether the increase was primarily in somata (perhaps reflecting increased biosynthesis), dendrites (reflecting increased transport), or at the postsynaptic membrane (reflecting functional glutamate receptors). At the EM, structural details were clearly visible even in the absence of osmium, allowing identification of glomerular terminals at the end of PA fibers. Myelin whorls and glycogen particles were observed on the lesioned side, but not on the control side (Kapadia and LaMotte 1987; Zhang X et al. 1995a). Another change apparent on the lesioned side involved glomerular terminals that in control material have dark axoplasm, few mitochondria and clear vesicles of irregular size. These terminals correspond to the central element of type C1 glomeruli (Figs. 1, 3A; Ribeiro-da-Silva and Coimbra 1982, 1984). After peripheral nerve lesion, these terminals can no longer be identified (Castro-Lopes et al. 1990). Other glomerular terminals in superficial DH with clear axoplasm, numerous mitochondria, and clear vesicles of regular size, corresponding to the central element of type C2 glomeruli (Figs. 1, 19C,D; Ribeiro-da-Silva and Coimbra 1982), are likely to originate from small-caliber myelinated PA fibers. Quantitative analysis was performed on these terminals, since they were recognizable on the operated side as easily as on the control side. A larger number of particles coding for AMPA receptor subunits was evident at glomerular synapses on the lesioned (Fig. 19C) as compared to the control side (Fig. 19D). To verify these qualitative observations, we counted gold particles at synapses made by C2 terminals on the two sides in the three animals used for EM. In each of the animals, labeling at synapses of C2 terminals was significantly increased on the injured side, with ratios ranging from 1.35 to 1.72. A slight (7%–8%) increase in the length of the synaptic active zone may have contributed to this increase, but most of the increased labeling could be attributed to increased receptor density, as indicated by the density of gold particles per micrometer of synaptic contact. Nonparametric analysis confirmed that receptor density was significantly elevated on the injured side ($p \leq 0.01$, Mann-Whitney U-test). These data established AMPA receptor up-regulation at synapses of PAs ipsilateral to the lesion in each of the animals studied. Might this result arise from intra-animal variability? To address this issue, we further analyzed the data with a paired t-test, comparing the mean number of gold particles/synapse on the lesioned and unlesioned sides for the three animals. Notwithstanding inevitable variations in tissue processing, the mean labeling on lesioned and control sides for each animal was very consistent in our material, thus making it possible to reject the null hypothesis that the observed effect might arise from random variations among animals ($p>0.05$, two-sided t-statistic).

We took advantage of the characteristic morphology of different types of synapses in superficial laminae to address whether changes in glutamate receptors after peripheral injury are confined to synapses of PAs. Besides glomerular terminals, superficial laminae contain nonglomerular, dome-shaped terminals filled with clear, round vesicles, and making single asymmetric synaptic contacts. Most of these are glutamatergic terminals originating from interneurons or descending fibers (Rustioni and Weinberg 1989). We counted gold particles associated with synapses made by dome-shaped terminals (Figs. 19A,B) randomly selected from lamina II in the same grids used for counts of synapses at C2 terminals. The mean number of gold particles was not significantly changed: synapses made by dome-shaped terminals on the injured side had an average of 0.94 times as much labeling as synapses made by dome-shaped terminals on the control side. These results imply that the increase in GluR2/3 is selective for terminals of PAs.

Considerations The effects of nerve injury upon the first synaptic link in the SC have been studied in many experimental models, and reported in a vast literature. The reaction to peripheral injury consists in part of trophic changes related to attempts at regeneration (Sebert and Shooter 1993; Hökfelt et al. 1994); however, the altered sensations associated with injury are likely to involve neurotransmitter mechanisms. The present results are of special interest, as glutamate is the main transmitter released at synaptic sites of PA terminals in the spinal DH (Jessell et al. 1986; Valtschanoff et al. 1994). Relatively little information from microscopic evidence has been published on glutamate and its receptors after peripheral nerve injury. A modest increase in immunocytochemical staining for glutamate has been reported in the DH, 7–14 days after chronic constriction injury of the sciatic nerve (Al-Ghoul et al. 1993). This is in contrast with the decrease in staining, after the same type of injury or after nerve section, of neuropeptides released by PA fibers, e.g., substance P and CGRP (Bennett et al. 1989; Al-Ghoul et al. 1993; Hökfelt et al. 1994; Kajander and Xu 1995). LM evidence suggests that neuropeptide receptors are up-regulated in the postsynaptic target after peripheral injury (Schäfer et al. 1993; McCarson and Krause 1994; Croul et al. 1995; Abbadie et al. 1996), whereas the literature on glutamate receptors is equivocal (Pellegrini-Giampietro et al. 1994; Croul et al. 1995; Kus et al. 1995; Harris et al. 1996; LaMotte et al. 1996). A modest increase in mean LM staining was observed in the present study; image analysis revealed more substantial increases within strongly fluorescent spots. Some of these were somata, perhaps reflecting increased biosynthesis, and others were within the neuropil, suggesting increased staining at synapses. The latter possibility was confirmed by our EM evidence that peripheral nerve injury induced an increase in the number of glutamate receptors at synapses of small-caliber PAs terminating in the substantia gelatinosa. Because negative synapses were not included, it may be argued that the increased counts of gold particles shown here may have resulted in one or two gold particles at synapses that might otherwise be negative on the side of the lesion. As the results, however, demonstrate increased counts in strongly immunopositive synapses, the exclusion of negative synapses from the counts would be expected to reduce rather than increase the difference in gold particle counts between the control and operated sides. Though our data indicate that the increase was mainly in receptor density, we also detected a modest increase in active zone length. Even if this increase was confirmed in a larger group of animals, both increased length and density would lead to a greater number of postsynaptic receptors. The results are unlikely to reflect selective survival of those synapses that normally express receptors at high density, because, at variance with C1 terminals, we did not see signs of loss of C2 terminals. We chose to use material stained for an antibody that recognizes both GluR2 and GluR3, because this antibody gives intense staining in lamina II. Moreover, C2 terminals, clearly recognizable by their morphology and well preserved after peripheral injury, have a distinct affinity to label with GluR2/3 (Popratiloff et al. 1996a). In the superficial laminae, the antibody for GluR2/3 appears to stain primarily the GluR2 subunit (Popratiloff et al. 1996a), since the GluR3 subunit is only sparsely present there (Furuyama et al. 1993; Henley et al. 1993; Tölle et al. 1993, 1995; Pellegrini-Giampietro et al. 1994). LM evidence suggests that peripheral injury may result in up-regulation of GluR1 in superficial laminae of the DH (Harris et al. 1996; Popratiloff et al. 1996b). However, changes in glutamate receptors may be limited to those of the AMPA type. Binding density for ligands selective for NMDA receptors was virtually unchanged after dorsal rhizotomy (Croul et al. 1995). Although we show here that glutamate receptors at synapses of intrinsic origin in superficial laminae are not up-regulated, we cannot exclude changes in receptors at terminals that are not glutamatergic. For instance, the number of neurons that are immunopositive for GABA in the DH decreases 2 weeks after sciatic nerve section. Concomitant changes in GABA receptors may occur at synapses upon PA terminals (Castro-Lopes et al. 1993, 1995). Increased efficacy of excitatory synaptic transmission is supported by an increase in synaptic field potentials recorded from the DH ipsilateral to peripheral neuropathy (Colvin et al. 1996), and the similarity in the time course of hyperalgesia and up-regulation of AMPA receptors in the superficial laminae after chronic constriction injury (Harris et al. 1996). Up-regulation of AMPA receptors has been implicated as a mechanism for increased synaptic efficacy (Maren et al. 1993; Isaac et al. 1995). The present report provides the first direct evidence for increased receptor protein at the synapse. Because AMPA receptors in substantia gelatinosa are mainly of the high-efficiency "flip" type (Tölle et al. 1995), an increase in their concentration postsynaptic to PA terminals may be the main mechanism available for increasing synaptic efficacy in this system. This increase may contribute to central sensitization and neuropathic pain in humans.

This monograph also shows up-regulation of AMPA receptor proteins at the synapses. The concentration increase at primary afferent synapses is presumably the explanation for increased synaptic efficacy in this system.

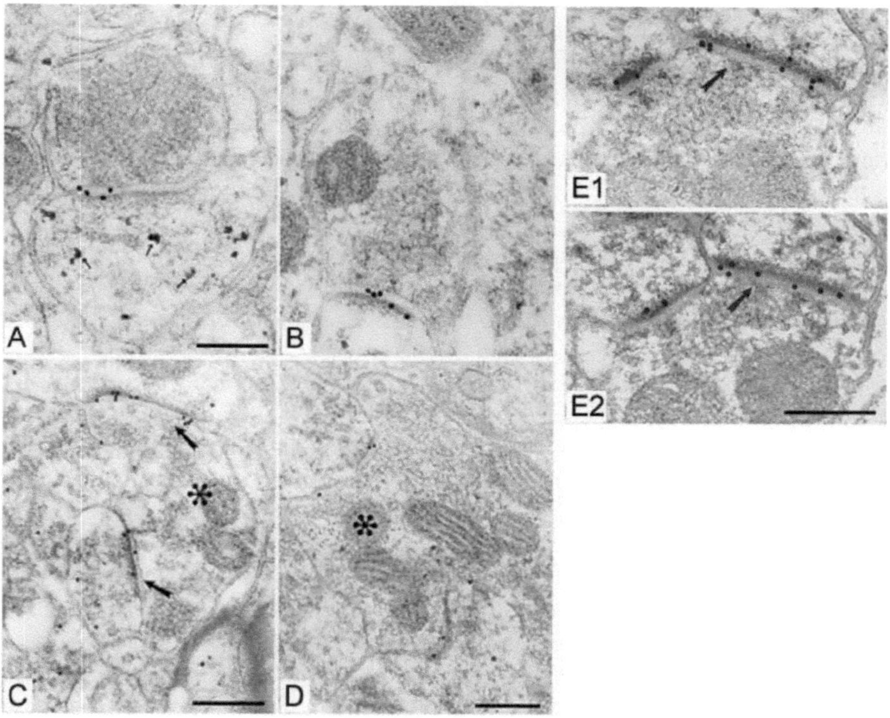

Fig. 19 A–E AMPA receptor subunits GluR2/3 labeling in PA terminals of the substantia gelatinosa of rat DH 2 weeks after sciatic nerve section. On the operated side, average labeling density does not change in the active zones sampled from small terminals likely originating from neurons in the spinal cord **A** compared to control side **B**. Average number of gold particles coding GluR2/3 increases at the active zones of C2 glomeruli, which is due to more frequently observed active zones with more gold particles on operated side **E1**, **E2**, and **C**, compared to the control side **D**. *Small arrows*—glycogen granules on injured side, *arrows*—strongly labeled active zones. Scale bars: **A** and **B**, 200 nm; **C**, 300 nm; **D**, **E2** and **E1**, 250 nm (Adapted with permission from Popratiloff et al. 1998a)

3.2 The Role of Glial Cells

There is growing evidence that both in the periphery and in the CNS, the glial cells play a modulatory role in the response to inflammation and injury, and in processes modifying nociception (Millan 1999; Watkins et al. 2001, 2003; Wieseler-Frank et al. 2004). The microglia is activated by NP (Eriksson et al. 1993; Honore et al. 2000; Tsuda et al. 2005). The substances derived from glial cells exert autocrine and paracrine effects and are able to globally effect activity in the SC (Aldskogius and Kozlova 1998; Minghetti and Levi 1998, Milligan et al. 2003, 2004; Watkins et al. 2003; Verge et al. 2004). According to Watkins et al. (2001) and to Watkins and Maier (2002b), the findings on the glial function—particularly that the glia express characteristics in common with immune cells—suggest a new, dramatically different approach to pain control, as all clinical therapies are focused exclusively on altering neuronal, rather than glial function.

Glia came to attention following the observations of Garrison et al. (1991) that following constriction of the sciatic nerves, the astrocytes in the lumbar SC display an increased staining for glial fibrillary acidic protein. The microgliocytes and astrocytes in the DH are presently known to show up-regulated expression of activation markers in response to different conditions that produce hyperalgesia, such as injury of the peripheral nerve (Colburn et al. 1999; DeLeo and Colburn 1999; Hashizume et al. 2000; Stuesse et al. 2000; Inoue et al. 2004; Ji and Strichartz 2004; Tsuda et al. 2005), subcutaneous formalin injection (Fu et al. 1999), experimental peritonitis (Watkins and Maier 2002a), experimental bone cancer (Schwei et al. 1999), SC injury (Popovich et al. 1997), and immune SC activation (Milligan et al. 2001). Increases in NGF and BDNF mRNA occur in Schwann cells and satellite cells in SG during inflammation of peripheral tissues (Cho et al. 1997), suggesting that by altering the expression and release of trophic factors, the Schwann cells and SG satellite cells may modulate nociceptive signaling. Peripheral axotomy induces a significant increase in NGF mRNA in the SG satellite cells, enhancing the pathologic sympathetic sprouting (Zhou et al. 1999). Other satellite cell-derived substances that might have demonstrable effects consistent with enhanced pain include glial cell-derived neurotrophic factor (GDNF), BDNF, neurotrophin-3, and proinflammatory cytokines (Watkins and Maier 2002a). The endothelins are peptides that have a diverse array of functions mediated by two receptor subtypes, the endothelin A and B receptors (Pomonis et al. 2001). Endothelin A receptor expression may play a role in signaling acute pain or NP, whereas endothelin B receptor expression may be involved in the transmission of chronic inflammatory pain. Pomonis et al. (2001) found the A receptor in a subset of small PA cells; however, the endothelin B receptor was not seen in the PA neurons but rather in the satellite cells and in nonmyelinating Schwann cells. These data indicate that the endothelins can have direct, nociceptive effects on the peripheral sensory nervous system and that peripheral glia may be directly involved in signaling nociceptive events in peripheral tissues. Madiai et al. (2002) examined the expression of fibroblast growth factor-2 (FGF-2) following ligation of lumbar spinal nerves. They found that FGF-2 was up-regulated both in PA neurons and in the SC astrocytes, suggesting neurotrophic functions of this growth factor following peripheral nerve lesion and possibly in astrocyte-related maintenance of pain states.

Watkins et al. (2001) recall the strange observation in the AIDS clinical literature that most patients suffer from chronic pain, a high percentage of which is of unknown bodily origin. This suggests that spinal viral invasion, causing glial activation and proinflammatory cytokine release, might potentially explain such pain.

3.3 Neuropathology of Herpes Zoster and of Postherpetic Neuralgia

Varicella-zoster virus (VZV) is an alpha herpes virus that is found exclusively in humans. VZV can cause a wide spectrum of disorders throughout life (Gilden et al. 2000; Kleinschmidt-DeMasters and Gilden 2001). This highly contagious virus causes a relatively benign disease in childhood: varicella (chickenpox). CNS complications are estimated to occur in less than 1% of chickenpox cases, and even this low number may be an overestimate (Kleinschmidt-DeMasters and Gilden 2001). Children have mild meningitic symptoms. The most common abnormality is cerebellar ataxia; very rarely, transverse myelitis has been reported.

After varicella resolves, VZV becomes latent in the SG and in the sensory ganglia of the cranial nerves and persists throughout the life of the host (Esiri and Tomlinson 1972; Gilden et al. 1983, 1987, 2000; Hyman et al. 1983; Croen et al. 1988; Mahalingam et al. 1990, 1999; Dueland et al. 1995; Esiri and Kennedy 1997; Kennedy et al. 1998; Cohrs et al. 2000; Kleinschmidt-DeMasters and Gilden 2001; Gilden et al. 2003). The data on the cellular localization of the latent VZV are contradictory. Croen et al. (1988) and Meier et al. (1993) opposed the common belief that VZV is localized in neurons, and declared that VZV is localized exclusively in the perineuronal satellite cells in latently infected human TG. Recent publications indicate that the virus is located predominantly in the pseudounipolar PA neurons, but the satellite cells are also implicated as a potential reservoir of latent VZV (Lungu et al. 1995, 1998; Kennedy et al. 1998; Mahalingam et al. 1999). During latency, VZV is not infectious and does not transcribe most of its genetic material, thereby escaping detection and clearance of the virus by the immune system.

The likelihood of viral reactivation to HZ increases with each advancing decade of age. HZ usually develops in elderly individuals and is eight to ten times more frequent after the age of 60 years than before (Kost and Straus 1996; Bowsher 1999c). Immunocompromised patients are at especially high risk (Kleinschmidt-DeMasters and Gilden 2001). With reactivation, the virus spreads transaxonally to the skin, causing a rash with a dermatomal distribution, and is associated with severe radicular pain. Any level of the neuraxis might be involved, but thoracic HZ is the most common one, affecting one to two, rarely more dermatomes, followed by the ophthalmic division of the 5th nerve (Hope-Simpson 1965; Portenoy et al. 1986; Zaal et al. 2000; Kleinschmidt-DeMasters and Gilden 2001; Devulder 2002).

HZ ophthalmicus may be associated with keratitis, a potential cause of blindness of the affected eye. The involvement of the facial nerve results in HZ oticus, often combined with paresis of the ipsilateral muscles of facial expression: geniculate neuralgia, described as early as 1907 by Ramsay Hunt (Hunt 1907, 1937; Brodal 1981). Similar combination of painful dermatomal rash with myotomal motor weakness might be observed also in the spinal nerve HZ (Yaszay et al. 2000). In the majority of patients, a prodrome of dermatomal pain starts before the appearance of the characteristic rash (Dworkin and Portenoy 1996; Dworkin and Johnson 1999). Dermatomal pain without a rash (zoster sine herpete) occurs rarely (Lewis 1958; Gilden et al. 1992). HZ is monophasic with recurrence occurring in less than 5% of immunocompetent patients. In contrast, in immunocompromised patients (especially in AIDS patients) HZ is recurrent, protracted, and often accompanied with severe neurological complications (De La Blanchardiere et al. 2000; Gilden et al. 2000, 2003).

The neuropathological investigation of HZ was started by the monograph of Head and Campbell (1900), reviewed by Oaklander (1999). Also quite early, von Bokay (1909) postulated an infectious agent common to varicella and HZ. The basic pathologic substrate for HZ is ganglionic hemorrhage, necrosis, and inflammation (Ghatak and Zimmerman 1973; Nagashima et al. 1975; Kleinschmidt-DeMasters and Gilden 2001). The histopathologic features include mononuclear and lymphocytic infiltration, neuronal degeneration, neuronal phagocytosis by satellite cells, empty neuronal cell beds, and fibrous scarring of the ganglia (Kleinschmidt-DeMasters et al. 1996; Esiri and Kennedy 1997). Vasculitis in the adjacent nerve results in damage of the axons (Gilden et al. 1996; Kleinschmidt-DeMasters et al. 1996), and especially destroyed are the myelin sheaths (Fabian et al. 1997). Rarely, VZV spreads in the CNS. The virus might spread both in centripetal and centrifugal directions (Schmidbauer et al. 1992; Kleinschmidt-DeMasters and Gilden 2001) causing myelitis (Hogan and Krigman 1973; Esiri and Kennedy 1997). In patients with HZ ophthalmicus, the virus might spread via trigeminal afferent fibers to the large blood vessels at the base of the brain, with resultant vessel thrombosis, vessel wall inflammation, and large, ipsilateral brain infarctions (Reshef et al. 1985; Gilden et al. 1996).

Most HZ in immunocompetent patients resolves without sequelae. However, many elderly patients have prolonged, debilitating pain, known as PHN. The increased incidence with increasing age is well known (Kost and Straus 1996; Bowsher 1999c; Dworkin and Johnson 1999; Helgason et al. 2000; Jung et al. 2004). The incidence of PHN has also been found to be much higher in adults with cancer (Lojeski and Stevens 2000) and in patients experiencing psychologic and physiologic stress (Livengood 2000). Jung et al. (2004) examined 965 HZ patients. They found out that older age, female sex, presence of a prodrome, greater rash severity, and greater acute pain severity made independent contributions to identifying which patients developed PHN.

According to Dworkin et al. (1997), five different types of pain may characterize PHN: throbbing pain, steady burning pain, intermittent sharp or shooting pain, allodynia, and hyperpathia (see below). Chronic pain causes suffering and distress. Here, pain became a disease itself, and it is a ruining disease (Portenoy et al. 1986; Nurmikko et al. 1991; Watson et al. 1991; Dworkin and Portenoy 1996; Attal and Bouhassira 1999; Dworkin and Johnson 1999; Dworkin et al. 2000; Gilden et al. 2000; Kanazi et al. 2000; Dworkin 2002). Dworkin and Johnson (1999) start their handbook article with an impressive phrase: The Norwegians have an admirable name for zoster (which like shingles means belt): "a belt of roses from hell", while the Danes call it "hell-fire."

Unfortunately, pharmacotherapy of NP is limited. Patients with PHN do not respond to nonsteroidal and anti-inflammatory drugs, and resistance or insensitivity to opiates is common (Bowsher 1997; Ossipov et al. 2000; Kanazi et al. 2000; Panlilio et al. 2002; Dworkin and Schmader 2003; Pappagallo and Haldey 2003; Harden 2005). Recent research (Panlilio et al. 2002; Dworkin and Schmader 2003; Lilie and Wassilew 2003) has shown that antiviral therapy can significantly reduce the risk and duration of postherpetic neuralgia in elderly patients, provided that treatment is started early in the course of disease (Jung et al. 2004).

The pathology of PHN is just beginning to be understood, and much less morphologic information is available for this condition than for HZ (Kleinschmidt-DeMasters and Gilden 2001). Along with the investigation of human material (Smith 1978; Watson et al. 1991; Rowbotham and Fields 1996; Rowbotham et al. 1996; Oaklander et al. 1998; Gilden et al. 2003), more or less successful animal models of NP conditions were developed (Willis et al. 1995; Attal and Bouhassira 1999; Honore et al. 2000b). Smith (1978), utilizing both LM and EM, described cystic distortion of thoracic SG removed 2.5 months after the onset of HZ, and persistent chronic inflammatory cells. He found "ghost cells" in a patient with removed SG 2 years after the onset of PHN, and hypothesized that the altered structure of surviving cells might contribute to the pathophysiology of the intractable pain. Watson et al. (1991) reported findings in the SG and adjacent portions of the nerve and of the rootlets in three cases with severe PHN and in two cases with no persistent pain. The findings of DH atrophy and cell, axon, and myelin loss were encountered only in patients with persistent pain. Marked loss of myelin and axons in the nerve and/or sensory roots were found in cases with and without pain. Not unexpectedly, Rowbotham et al. (1996) and Oaklander et al. (1998) demonstrated a greater loss of small cutaneous nerve endings in skin biopsies obtained from patients with HZ who developed PHN than in those who developed no neurologic sequelae.

3.4 Diabetic Neuropathic Pain

The diabetic neuropathy is a severe late complication of diabetes mellitus, and is the most common cause of neuropathy in the Western world (Simmons and Feldman 2002). Its pathogenesis is multifactorial, involving both metabolic and vascular factors (Feldman et al. 1999; Eaton et al. 2001). Diabetic neuropathy has been extensively studied in experimental animals exposed to the hyperglycemic agent streptozocin (Fox et al. 1999). The NP involves predominantly the distal portions of the extremities (Vrethem et al. 2002). It has been suggested that diabetic NP results from hyperactivity of damaged C-fibers (Chen and Levine 2001; Kapur 2003; McHugh and McHugh 2004). In addition, the electrophysiological data of Khan et al. (2002) provide evidence that an abnormal sensory input not only from C- and Aδ-fibers, but also from Aβ-fibers may play an important role in diabetic NP. Heavy alterations of the myelinated axons (onion-bulb formation) in patients with diabetic neuropathy were first described by Thomas and Lascelles (1966). Severe damage of the myelin sheaths in the dorsal and ventral lumbar roots of rats after 8 months of streptozotocin-induced diabetes was reported by Tamura and Parry (1994). Mizisin et al. (1998) examined biopsies from cats with spontaneously occurring diabetes with the electron microscope, and Kalichman et al. (1998) observed biopsy samples from the sural nerve of patients with diabetic neuropathy. In both studies, the most evident finding was a heavy myelin defect characterized by splitting and ballooning of the sheath, while the axons were relatively spared. Schwann cell injury was significant. The reactive changes included accumulations of Pi granules of Reich, lipid droplets and intermediate cytoplasmic filaments. Degenerative changes ranged from dissolution of Schwann cell cytoplasm at the inner glial loop associated with periaxonal swelling and axonal shrinkage to demyelination. According to Eckersley (2002), hypoxia, hyperglycemia, and increased oxidative stress contribute directly or indirectly to Schwann cell dysfunction in diabetic neuropathy. The results include impaired paranodal barrier function, damaged myelin sheaths, reduced antioxidative capacity, and decreased neurotrophic support for axons.

There are few data on the central mechanisms of diabetic NP, although DeJong (1977) found that lesions of the SC are not uncommon and may result in pain syndromes.

3.5 Cancer Neuropathic Pain

For many patients, pain is the first sign of cancer, and 30%–50% of all cancer patients will experience moderate to severe pain; the frequency and intensity of pain tend to increase during the advanced stages, so that 75%–95% of patients with metastatic or advanced-stage cancer will experience severe pain (Portenoy 1992; Portenoy et al. 1999; Regan and Peng 2000; Mantyh et al. 2002). In the cancer population, NP is often related to compression, direct neoplastic invasion of the peripheral nerves and/or the SC, or to a neuropathy caused by chemotherapy (Farrar and Portenoy 2001). Manfredi et al. (2003) examined 187 patients with cancer and pain, and the pain was categorized as neuropathic in 103 patients. The most frequent sites of neurological injury were nerve roots, SC and cauda equina, brachial and lumbosacral plexus, and peripheral nerves. There were no patients with pain caused by injury to the brain.

Strangely enough, although not significantly, some tumors might be innervated by sensory neurons (O'Connell et al. 1998; Seifert and Spitznas 2001; Terada and Matsunaga 2001), but this is not the main reason for the cancer pain. Far more importantly, the tumor frequently entraps and injures the nerves by compression, ischemia, or proteolysis. The proteolytic enzymes injure the sensory and sympathetic nerve fibers, causing NP (Mantyh et al. 2002). Along the tumor cells, the tumor contains immune-system cells (macrophages, neutrophils, T cells). Both tumor and inflammatory cells secrete numerous factors that sensitize or directly excite PA neurons: prostaglandins, tumor necrosis factor-α, endothelins, interleukin-1 and -6, epidermal growth factor, transforming growth factor-β, and platelet-derived growth factor (Mantyh et al. 2002).

The most comprehensive studies on cancer NP concern bone cancer (Mantyh et al. 2002; Clohisy and Mantyh 2003; Sabino et al. 2003, 2005), since there is a reliable experimental animal model (Honore et al. 2000a, b; Mantyh and Hunt 2004). It appears that bone cancer pain represents a unique pain state (Clohisy and Mantyh 2003). In the persistent inflammatory pain state induced by subcutaneous injection of capsaicin, there is besides massive functional and electrophysiological changes an up-regulation of SP and CGRP, while a down-regulation of these neuropeptides takes place in the NP state following nerve transection or ligation (Noguchi et al. 1989; Villar et al. 1991; Donnerer et al. 1993; Garrison et al. 1993; Safieh-Garabedian et al. 1995; Cho et al. 1996, 1997). However, Honore et al. (2000b) found no significant changes in the expression of these neurotransmitters in the murine bone cancer model. These authors encountered large differences that occur with each pain state in the SC: inflammation induced an increase in SP and CGRP in laminae I and II, neuropathy induced down-regulation of these transmitters, while bone cancer had no effect. In the murine model of bone cancer pain, Schwei et al. (1999) observed a dramatic up-regulation of glial fibrillary acidic protein in the SC, indicating massive astrogliosis. Again, this phenomenon is not observed in the inflammatory and neuropathic pain models.

3.6 Central Neuropathic Pain

3.6.1 Spinal Cord Injury

Chronic NP occurs in approximately 50% (varying from 42% to 77%) of patients with SC injury (Bonica 1991; Anke et al. 1995; Levi et al. 1995; Eide et al. 1996; Christensen and Hulsebusch 1997; Bowsher 1999a; Siddall et al. 1999; Siddall and Loeser 2001; Finnerup et al. 2001; Finnerup and Jensen 2004). Syringomyelia is a rare disease but with a very high incidence of central pain (Boivie 1999). In 22 patients, he found that all had pain.

There are two varieties of NP following SC injury: (a) at-level pain (in segments corresponding to the level of SC injury), (b) below-level pain (in parts of the body corresponding to segments caudal to the injury) (Siddal et al. 2000). According to Siddal et al. (1997), the below-level NP should be considered as a central pain condition caused by the SC injury, while at-level pain may have peripheral and central components that are difficult to separate.

Eide et al. (1996) compared somatosensory abnormalities in painful and non-painful denervated areas at or below injury in patients with SC injuries. They observed that allodynia was more common in painful areas, and suggested that in the pathogenesis of NP a major role is played by the hyperexcitability of STT neurons. Bouhassira et al. (2000) investigated patients with painful and painless syringomyelia. They observed no significant differences in thermal or mechanical sensory function between patients with or without pain. Similarly, Defrin et al. (2001) found no differences between thermal and tactile sensations in patients with or without pain, but allodynia was only elicited in pain patients. Finnerup et al. (2003) compared clinical examination, quantitative sensory testing and somatosensory evoked potentials in patients with traumatic SC injury with and without pain below spinal lesion level. The patients with central pain more frequently had sensory hypersensitivity in dermatomes corresponding to the level of the injury. They found a significant correlation between the disesthesia at the level of the lesion and spontaneous pain caudal to the injury level. Finnerup et al. (2003) think that STT damage may be a necessary but not sufficient condition for developing below the level pain, since deficits of STT functions were equally severely affected in patients without pain. They propose that pain in body segments below the level of injury should be linked to the presence of abnormal evoked sensations in segments at the level of injury. Finnerup et al. (2003) suggest that neuronal hyperexcitability in second- or third-order neurons, which have lost their normal afferent input, is an important mechanism for pain below spinal injury.

3.6.2 Brain Injury

According to Boivie (1999), historically central pain appears to have been first described as early as 1883 by Greiff in a patient who, following cerebrovascular lesions including the thalamus, developed *reissende Schmerzen* (tearing pain). The term "thalamic syndrome" was introduced by Dejerine and Roussy (1906), who described three cases of a condition in which spontaneous pain followed a stroke, and the autopsies showed the infarct to be in the thalamus. Presently, the condition is known as central post-stroke pain (Leijon et al. 1989; Bowsher 1999b). Vestergaard et al. (1995) reported that approximately 8% of all stroke patients develop central post-stroke pain. Lesions at any level along the neuraxis can cause central pain. Thus lesions at the first synapse in the DH of the SC or trigeminal nuclei, along the ascending pathways, in the thalamus, in the subcortical white and probably in the cerebral cortex have all been reported to cause central pain (Riddoch 1938; Garcin 1968; Cassinari and Pagni 1969; Leijon et al. 1989; Tasker 1990; Bowsher 1996, 1999a; Pagni 1998). The highest prevalence has been noticed after lesions of the SC, lower brainstem and ventroposterior part of the thalamus (Bonica 1991; Boivie 1992, 1999). The importance of the location of the thalamic lesion was repeatedly evaluated. According to Bogoussslavsky et al. (1988), only patients with lesions including the ventroposterior thalamus develop central pain. Also in studies carried out by Leijon et al. (1989) and Bowsher et al. (1998), all thalamic lesions included part of the ventroposterior thalamic nuclei. Boivie (1999) recalls that this is in accordance with Hassler's idea that the V.c.p.c. thalamic nucleus (an important site of STT termination in humans), is the crucial location for thalamic lesions causing central pain (Hassler 1960).

The lesions that cause central pain vary enormously in location, size, and structure. There is no study indicating that a small lesion in the DH of the SC carries less risk for central pain than a huge infarct involving much of the thalamus and large parts of the white matter lateral and superior to the thalamus (Boivie 1999). There are several hypotheses concerning the mechanisms involved in the pathophysiology of central pain. Head and Holmes (1911) proposed a disinhibition of the STT-thalamocortical system, triggered by lesions of the DCN-medial lemniscus system. Foerster's hypothesis (1936) was similar: he thought that epicritic sensitivity (touch, pressure, vibration) normally exerts control over protopathic sensitivity (pain and temperature). According to Foerster's hypothesis, central pain can only occur when there is a loss of epicritic sensitivity, e.g., destruction of the lemniscal system. More recently, indications were repeatedly found that the STT system is affected in the majority of patients with central pain (Boivie et al. 1989; Tasker 1990; Bowsher et al. 1998; Pagni 1998). Central pain patients have abnormal temperature and pain sensibility, but they may have normal threshold to joint movements, vibration, and touch (Boivie et al. 1989; Bowsher et al. 1998). Low brainstem infarcts (Wallenberg syndrome) and cordotomies, in which the STT but not the medial lemniscus is interrupted, may cause central pain (Boivie et al. 1989; Bowsher 1996; Bowsher et al. 1998; Pagni 1998; Boivie 1999). Probably, the crucial lesions affect the neo-STT, e.g., the projection to the ventroposterior thalamic region (Garcin 1968; Bowsher 1996; Pagni 1998). This kind of lesion leave the more medially and inferiorly terminating paleo-STT projections intact (Boivie 1999).

Another hypothesis focuses on the role of the reticular thalamic nucleus, and the medial and intralaminar zones receiving STT fibers. The reticular nucleus is the only thalamic nucleus built by GABAergic, inhibitory projection neurons that do not give rise to thalamocortical fibers but heavily innervate the remaining thalamic nuclei (Gonzalo-Ruiz and Lieberman 1995; Guillery et al. 1998; Jones 2002a, b; Guillery and Harting 2003). According to this hypothesis, the lesion removes the suppressing activity exerted by the reticular thalamic nucleus on intralaminar and medial thalamic nuclei, thereby releasing abnormal activity in this region, which in turn leads to pain and hypersensitivity (Cesaro et al. 1991). In accordance with this hypothesis, Edgar et al. (1993) pointed out that most thalamic lesions that cause central pain might involve part of the reticular nucleus, as well as parts of the ventroposterior nuclei.

Recently Craig (1998, 2003d) put forward a new hypothesis about the mechanisms of central pain. His hypothesis builds on the classical insights of Head and Holmes (1911). Central pain is due to the disruption of thermosensory integration and the loss of cold inhibition of burning pain. This disruption is caused by a lesion along the STT to the nuclei VPI, VMpo, and MDvc. These projections tonically inhibit nociceptive thalamocortical neurons, which by the lesion increase their firing and produce pain. The pathway is activated by cold receptors in the periphery, which in turn activate cold-specific and polymodal lamina I cells in the DH. According to Boivie (1999), this hypothesis might be applicable in some patients, but not in others, because of the location of the lesions and the character of the pain (roughly 40%–60% of all central pain has a burning character).

3.6.3 Changes in Cortical Networks Due to Chronic Pain

Persistent pain causes suffering and distress, and pain can become a disease in itself. Chronic pain or NP can result from damage of the nervous system at different levels of pain processing: peripheral nerve, SG, dorsal root, CNS. Chronic syndromes mostly show positive symptoms such as pain, dysesthesia, and paresthesia, often in combination with negative symptoms such as sensory deficits. Unfortunately, pharmacotherapy of NP is limited, perhaps because the etiology, the mechanisms, and the symptoms of NP may differ considerably between patients. In patients with PHN pain, mainly peripheral mechanisms are discussed as being involved, but central changes might also be involved. Peripheral neuropathic pain is a spontaneous pain (stimulus-independent pain) or a hypersensitivity pain caused by a stimulus following damage of sensory neurons (stimulus-evoked pain). Inflammation in the DG can sensitize neurons to respond to normal innocuous thermal or mechanical stimuli and loss of DG perikarya can induce changes in surrounding surviving neurons. Thus, loss of sensory dendrites in the epidermis of patients suffering from PHN was positively correlated with both sensory deficits and with pain (Baron and Saguer 1993; Koltzenburg et al. 1994; Woolf and Doubell 1994; Rowbotham et al. 1996; Oaklander et al. 1998). Changes caused by alterations of peripheral input, followed by altered spinal processing can be forwarded to the cortex via thalamic nuclei (Coderre et al. 1993; Hsieh et al. 1995). Neurons in the somatosensory thalamus of patients with NP showed various electrophysiological abnormalities: responses to stimuli of body regions not normally driving those cells, high spontaneous firing rates, and abnormal bursting activities. Thus, besides peripheral and spinal changes there is massive cortical plasticity contributing to the development of pathological pain.

Neuropathic pain has been studied using CT, EEG, PET, and fMRI. Substantial plastic changes were found in the cortex using these techniques. Functional reorganization in SI and SII were described. In NP, a change related to chronicity and amount of pain was reported (Flor 2003). Moreover, using ^{1}H-MRS chemical changes were noted in the dorsolateral prefrontal cortex in chronic back pain for N-acetyl aspartate and glucose, which could be related to measures of pain and anxiety (Grachev et al. 2000).

One of the thoroughly investigated topics in this field is the processing of ongoing phantom limb pain (Flor et al. 1995, 1997). In humans, it could be shown that there is a takeover of SI representation fields, no longer "used" because of limb amputation, by directly adjoining cortical fields representing adjacent areas of the body surface (Birbaumer et al. 1997): in SI the representation of the lower lip of the side of amputation was found in the position that should have been occupied by the contralateral (amputated) upper extremity. Changes in the cortical presentation are intimately correlated with phantom limb pain (Flor et al. 1995, Birbaumer et al. 1997): larger amounts of cortical reorganization are correlated with increased pain impression (Montoya et al. 1998). In patients suffering from pain due to traumatic upper limb loss, pharmacological blockade of the respective brachial plexus could reverse the "pathological" cortical map in those patients that showed a pain reduction (Birbaumer et al. 1997).

That chronic ongoing NP (painful mononeuropathy) altered cortical activity was shown by a positron emission topography study comparing patients' habitual pain state with that of a pain-alleviated state induced by regional nerve block with lidocaine (Hsieh et al. 1995). Although activities of SI and SII were not significantly altered during both states in the patients, there was a clear state difference in the activities of the IC, the posterior parietal, and the inferior lateral prefrontal cortices, indicating an involvement of those areas in NP processing. Most interestingly, the ACC of the right hemisphere was found activated irrespective of the body side of the painful nerve. The noninvolvement of SI in chronic pain corroborates the observation that surgical extirpation of SI and SII provided little or no relief from chronic pain. The cerebral activation pattern in neuropathic patients found by Hsieh et al. (1995) shows the importance of the motivational-affective dimension of pain in this illness. These findings are also in accordance with reports (Craig et al. 1996a) that the ACC is activated in models demonstrating illusion of pain.

4 Concluding Remarks

The neuroanatomy of pain is complex, as many ascending systems in parallel are involved in pain processing. Even more complex is the neuropathology of PHN as far as it is understood to date. Damage of the nervous system at the level of SG results in a rearrangement of the highly ordered laminar termination of PAs within the appropriate regions of the DH. Normally, unmyelinated C-fibers terminate in lamina II, myelinated mechanoceptive $A\beta$-axons in laminae III–VI of the SC. Following the virus-induced transganglionic degeneration of C-axons, long-lasting sprouting of A-fibers into lamina II occurs. The functional importance of A-fiber sprouting is that lamina II begins to receive information about non-noxious stimuli. This information may be misinterpreted by the CNS as noxious. Thus PHN can be interpreted firstly as a result of a massive sprouting on the level of the SC, secondly leading to abnormal ascending projection that thirdly are pathologically further processed in the brainstem, the thalamus, and the cortical areas involved in pain perception.

5 Summary

Pain is an unpleasant but very important biological signal for danger. Nociception is necessary for survival and maintaining the integrity of the organism in a potentially hostile environment. Pain is both a sensory experience and a perceptual metaphor for damage and it is activated by noxious stimuli that act on a complex pain sensory apparatus. However, chronic pain no longer having a protective role can become a ruining disease itself, termed neuropathic pain.

From periphery to cortex, the neuroanatomical chain of pain consists of the primary afferent (PA), the perikarya localized in spinal ganglia (SG) and in the sensory ganglia of the 5th, 7th, 9th and 10th nerves. The largest A cells are typical proprioceptor, and the small B cells are typical nociceptor neurons. The peripheral processes of the nociceptive PA cells are thin fibers of two types: Aδ- and C-fibers, the Aδ-fibers being responsible for the "first pain" (pinprick sensation), and C-fibers for the "second pain" (burning or dull pain). The free nerve endings are to be found throughout the body, mainly in the adventitia of small blood vessels, in outer and inner epithelia, in connective tissue capsules, and in the periosteum.

As central processes of SG neurons, the nociceptive fibers terminate primarily in laminae I and II; the Aδ-fibers terminate in laminae I and V, and C-fibers in laminae I and II. The nociceptive-specific neurons are dominated by Aδ-fiber input. The polymodal nociceptive cells are dominated by C-fiber input and are important for the second pain. The central processes of pseudounipolar TG neurons mostly descend especially to the caudal part of the spinal trigeminal nucleus, with a structure similar to the spinal dorsal horn. Two types of glomerular terminals could be identified in superficial laminae resembling terminals of unmyelinated or from thinly myelinated PAs. In the superficial laminae of the SC, especially glutamate receptors and their relation to types of synapses play a crucial role for decoding the convergent inputs at the level of the first brain synapse and for the understanding of abnormal pain.

A distribution of GluR1 and GluR2/3 for AMPA receptors is described in the superficial dorsal horn of the spinal cord. GluR1 showed a lateral localization, while GluR2 was localized over the mediolateral extent of the superficial dorsal horn. Electron microscopic results revealed that GluR1 antibody was related to C1 synapses, while GluR2/3 antibodies were localized on C2 synapses.

Ascending pathways of the spinal cord (SC) and of the spinal trigeminal nucleus, the spino- and the trigeminothalamic tracts, mediate the sensations of pain, cold, warmth, and touch. The cells of origin are located mainly in laminae I and IV–VI, their mostly crossing axons reaching various nuclei of the thalamus. Also, the dorsal column nuclei (DCN) are highly involved in nociception, projecting via the medial lemniscus to thalamic nuclei. Furthermore, the entire trigeminal sensory nuclear complex projects to the thalamus.

Our retrograde axonal transport studies revealed the projections to the ventrobasal thalamus in the rat. In the brainstem, the contralateral principal sensory and all subdivisions of the spinal trigeminal nucleus contained retrogradely labeled neurons, but to a different extent. The ipsilateral projection to the thalamus is faint but unquestionable. The experiments also demonstrated a prominent crossed connection from the DCN to the thalamus. Only a few neurons in the DCN ipsilateral to the injection were labeled. In the SC, the distribution of labeled neurons was uneven. The highest number of labeled neurons was encountered at the spinomedullary junction and in the four cranial cervical segments, mostly contralateral to the thalamic injection. In the more caudal segments, the number of labeled neurons decreased.

There is a multiregional organization of supraspinal pain processing, and the cortical areas involved in pain perception are the primary (SI) and secondary (SII) somatosensory cortex, the insular (IC), the anterior cingulate (ACC), and the prefrontal (PC) cortices. The sensory-discriminative aspect of pain (localization, intensity, duration, quality) is presented in SI and SII, the motivational-affective aspect (subjective suffering, unpleasantness, aversive emotions) and the cognitive-evaluative aspects of pain are presented in IC, ACC, and PC.

Pain processing is controlled by descending modulatory pathways. Neurons from the periaqueductal grey (PAG) project to the serotoninergic raphe nuclei of the medulla oblongata and to the noradrenergic nuclei in the dorsolateral pons. Both the catecholaminergic and indolaminergic neuronal groups project heavily to the SC and to the spinal trigeminal nucleus. Along with serotonin and noradrenaline, also endogenous opiates and the amino acids glutamate, GABA, and glycine are clearly involved. It was suggested that descending facilitatory influences could contribute to chronic pain states, and such influences were important to the development and maintenance of hyperalgesia.

Chronic, maladaptive neuropathic pain typically results from damage to the nervous system. Several etiologies of peripheral nerve injury might result in neuropathic pain: postherpetic neuralgia (PHN), traumatic injury, phantom limb pain, diabetes, and malignancy. Neuropathic pain conditions share certain clinical characteristics: spontaneous, continuous pain, usually of a burning character; paroxysmal (shooting, lancinating) pain; and evoked pain to various mechanical or thermal stimuli, such as allodynia and hyperalgesia.

As the result of neuroanatomical and neurochemical plasticity in the CNS, peripheral nerve injury results in transsynaptic degeneration and a rearrangement of the highly ordered laminar termination of PAs within appropriate regions of the dorsal horn.

Chronic neuropathic pain occurs in approximately 50% of patients with SC injury and following various brain injuries, i.e., central post-stroke pain in approximately 8% of all stroke patients and lesions at any level along the neuraxis. Chronic pain seems to be the result of changes in cortical networks.

PHN can be interpreted firstly as a result of a massive sprouting on the level of the SC, secondly leading to abnormal ascending projection that thirdly are pathologically further processed in the brainstem, the thalamus, and the cortical areas involved in pain perception.

Acknowledgements We gratefully thank Barbara Kuhnke (Rostock), Snejina S. Ilieva (Sofia), and Gergana Genova (Washington) for their excellent technical assistance. Research was partially supported by a grant from the National Science Fund of Bulgaria (L1012/2001).

References

Abbadie C, Brown JL, Mantyh PW, Basbaum AI (1996) Spinal cord substance P receptor immunoreactivity increases in both inflammatory and nerve injury models of persistent pain. Neuroscience 70:201–209

Al-Chaer ED, Lawand NB, Westlund KN, Willis WD (1996a) Visceral nociceptive input into the ventral posterolateral nucleus of the thalamus: a new function for the dorsal column pathway. J Neurophysiol 76:2661–2674

Al-Chaer ED, Lawand NB, Westlund KN, Willis WD (1996b) Pelvic visceral input into the nucleus gracilis is largely mediated by the postsynaptic column pathway. J Neurophysiol 76:2675–2690

Al-Chaer ED, Westlund KN, Willis WD (1997) Nucleus gracilis: an integrator for visceral and somatic information. J Neurophysiol 78:521–527

Al-Chaer ED, Feng Y, Willis WD (1998) A role for the dorsal column in nociceptive visceral input into the thalamus of primates. J Neurophysiol 79:3143–3150

Aldskogius H, Kozlova EN (1998) Central neuron–glial and glial–glial interactions following axon injury. Prog Neurobiol 55:1–26

Aley KO, Levine JD (2002) Different peripheral mechanisms mediate enhanced nociception in metabolic/toxic and traumatic painful peripheral neuropathies in the rat. Neuroscience 111:389–397

Al-Ghoul WM, Li Volsi G, Weinberg RJ, Rustioni A (1993) Glutamate immunocytochemistry in the dorsal horn after injury or stimulation of the sciatic nerve of rats. Brain Res Bull 30:453–459

Allen GV, Barbrick B, Esser MJ (1996) Trigeminal-parabrachial connections: possible pathway for nociception-induced cardiovascular reflex responses. Brain Res 715:125–135

Alles A, Dom RM (1985) Peripheral sensory nerve fibers that dichotomize to supply the brachium and the pericardium in the rat; a possible morphological explanation for referred cardiac pain? Brain Res 342:382–385

Alm P, Uvelius B, Ekstrom J, Holmqvist B, Larsson B, Andersson KE (1995) Nitric oxide synthase-containing neurons in rat parasympathetic, sympathetic and sensory ganglia: a comparative study. Histochem J 27:819–831

Alvarez FJ, Kavookjan AM, Light AR (1992) Synaptic interactions between GABA-immunoreactive profiles and the terminals of functionally defined myelinated nociceptors in the monkey and cat spinal cord. J Neurosci 12:2901–2917

Alvarez FJ, Kavookjan AM, Light AR (1993) Ultrastructural morphology, synaptic relationships, and CGRP immunoreactivity of physiologically identified C-fiber terminals in the monkey spinal cord. J Comp Neurol 329:472–490

Alvarez FJ, Harrington D, Fyffe REW (1994) AMPA and NMDA receptor-immunoreactivity post-synaptic to primary afferent terminals in the superficial dorsal horn of the cat spinal cord. Soc Neurosci Abstr 20:1570

Amassian VE (1951) Fiber groups and spinal pathways of cortically represented visceral afferents. J Neurophysiol 14:445–460

Ambalavanar R, Moritani M, Haines A, Hilton T, Dessem D (2003) Chemical phenotypes of muscle and cutaneous afferent neurons in the rat trigeminal ganglion. J Comp Neurol 460:167–179

Andres KH, von Düring M (1973) Morphology of cutaneous receptors. In: Iggo A (ed) Handbook of sensory physiology, vol II. Somatosensory system. Springer, New York, pp 1–28

Andres KH, von Düring M, Schmidt RF (1985) Sensory innervation of the Achilles tendon by group III and IV afferent fibers. Anat Embryol 172:145–156

Andrew D, Craig AD (2001) Spinothalamic lamina I neurons selectively sensitive to histamine: a central neural pathway for itch. Nature Neurosci 4:72–77

Andrew D, Craig AD (2002) Quantitative responses of spinothalamic lamina I neurones to graded mechanical stimulation in the cat. J Physiol 545:913–931

Anke AG, Stenehejm AE, Stanghelle JK (1995) Pain and life quality within 2 years of spinal cord injury. Paraplegia 33:555–559

Aoki E, Takeuchi IK, Shoji R, Semba R (1993) Localization of nitric oxide-related substances in the peripheral nervous tissues. Brain Res 620:142–145

Apkarian AV, Hodge CJ (1989a) Primate spinothalamic pathways. I. A quantitative study of the cells of origin of the spinothalamic pathway. J Comp Neurol 288:447–473

Apkarian AV, Hodge CJ (1989b) Primate spinothalamic pathways. II. The cells of origin of the dorsolateral and ventral spinothalamic pathways. J Comp Neurol 288:474–492

Apkarian AV, Hodge CJ (1989c) Primate spinothalamic pathways. III. Thalamic terminations of the dorsolateral and ventral spinothalamic pathways. J Comp Neurol 288:493–511

Apkarian AV, Brüggemann J, Shi T, Airapetian LR (1995) A thalamic model for true and referred visceral pain. In: Gebhart GF (ed) Visceral pain. IASP, Seattle, pp 217–259

Arvidsson J (1982) Somatotopic organization of vibrissae afferents in the trigeminal sensory nuclei of the rat studied by transganglionic transport of HRP. J Comp Neurol 211:84–92

Aston-Jones G, Shipley MT, Grzanna R (1995) The locus coeruleus, A5 and A7 noradrenergic cell groups. In: Paxinos G (ed) The rat nervous system. Academic Press, San Diego, pp 183–213

Attal N, Bouhassira D (1999) Mechanisms of pain in peripheral neuropathy. Acta Neurol Scand 173 (Suppl):12–24

Avendano C, Lagares A (1996) A stereological analysis of the numerical distribution of neurons in dorsal root ganglia C4-T2 in adult macaque monkeys. Somatosens Mot Res 13:59–66

Azkue JJ, Mateos JM, Elezgarai I, Benitez R, Lazaro E, Streit P, Grandes P (1998) Glutamate-like immunoreactivity in ascending spinofugal afferents to the rat periaqueductal gray. Brain Res 790:74–81

Bajic D, Proudfit HK (1999) Projections of neurons in the periaqueductal gray to pontine and medullary catecholamine cell groups involved in modulation of nociception. J Comp Neurol 405:359–379

Ball MJ, Nuttall K, Warren KG (1982) Neuronal and lymphocytic populations in human trigeminal ganglia: Implications for ageing and for latent virus. Neuropathol Appl Neurobiol 8:177–187

Barnett EM, Evans GD, Sun N, Perlman S, Cassell MD (1995) Anterograde tracing of trigeminal afferent pathways from murine tooth pulp to cortex using herpes simplex virus type 1. J Neurosci 15:2972–2984

Baron R, Saguer M (1993) Postherpetic neuralgia. Are C-nociceptors involved in signalling and maintenance of tactile allodynia? Brain 116:1477–1496

Baron R, Levine JD, Fields HL (1999) Causalgia and reflex sympathetic dystrophy: does the sympathetic nervous system contribute to the generation of pain? Muscle Nerve 22:678–695

Basbaum AI (1999) Spinal mechanisms of acute and persistent pain. Reg Anesth Pain Med 24:59–67

Basbaum AI, Fields HL (1984) Endogenous pain control system: brainstem spinal pathways and endorphin circuitry. Annu Rev Neurosci 7:309–338

Basbaum AI, Jessel TM (2000) The perception of pain. In: Kandel ER, Schwartz JH, Jessel TM (eds) Principles of neural science. McGraw-Hill, New York, pp 472–491

Bassetti C, Bogousslavsky J, Regli F (1993) Sensory syndromes in parietal stroke. Neurology 43:1942–1949

Battaglia G, Rustioni A (1992) Substance P innervation of the rat and cat thalamus. II. Cells of origin in the spinal cord. J Comp Neurol 315:473–486

Battaglia G, Spreafico R, Rustioni A (1992) Substance P innervation of the rat and cat thalamus. I. Distribution and relation to ascending spinal pathways. J Comp Neurol 315:457–472

Beggs J, Jordan S, Ericson AC, Blomqvist A, Craig AD (2003) Synaptology of trigemino- and spinothalamic lamina I terminations in the posterior ventral medial nucleus of the macaque. J Comp Neurol 459:334–354

Beitz AJ (1995) Periaqueductal gray. In: Paxinos G (ed) The rat nervous system. Academic Press, San Diego, pp 173–182

Belmonte C, Gallar J (1996) Corneal nociceptors. In: Belmonte C, Cervero F (eds) Neurobiology of nociceptors. Oxford University Press, Oxford, pp 146–183

Bennett GJ (1994) Neuropathic pain. In: Wall PD, Melzack R (eds) Textbook of pain. Churchill Livingstone, Edinburgh, pp 201–224

Bennett GJ, Xie YK (1988) A peripheral mononeuropathy in rats that produces disorders of pain sensation like those seen in man. Pain 33:87–107

Bennett GJ, Kajander KC, Sahara Y, Iadarola MJ, Sugimoto T (1989) Neurochemical and anatomical changes in the dorsal horn of rats with an experimental painful peripheral neuropathy. In: Cervero F, Bennett GJ, Headley PM (eds) Processing of sensory information in the superficial dorsal horn of the spinal cord. Plenum Press, New York, pp 463–471

Berkley KJ, Hubscher CH (1995) Are there separate central nervous system pathways for touch and pain? Nature Med 1:766–773

Berkley KJ, Blomqvist A, Pelt A, Flink R (1980) Differences in the collateralization of neuronal projections from the dorsal column nucleus and lateral cervical nucleus to the thalamus and the tectum in the cat: an anatomical study using two different double labelling techniques. Brain Res 202:273–290

Berkley KJ, Budell RJ, Blomqvist A, Bull M (1986) Output systems of the dorsal column nuclei in the cat. Brain Res Rev 11:199–225

Berkley KJ, Hubscher CH, Wall PD (1993) Neuronal responses to stimulation of the cervix, uterus, colon, and skin in the rat spinal cord. J Neurophysiol 69:545–556

Bernard JF, Besson JM (1990) The spino(trigemino)pontoamygdaloid pathway: electrophysiological evidence for an involvement in pain processes. J Neurophysiol 63:473–490

Bernard JF, Villanueva L, Carroue J, Le Bars D (1990) Efferent projections from the subnucleus reticularis dorsalis (SRD): a Phaseolus vulgaris leucoagglutinin study in the rat. Neurosci Lett 116:257–262

Bernard JF, Dallel R, Raboisson P, Villanueva L, Le Bars D (1995) Organization of the efferent projections from the spinal cervical enlargement to the parabrachial area and the periaqueductal gray: a PHA-L study in the rat. J Comp Neurol 353:480–505

Bernard JF, Bester H, Besson JM (1996) Involvement of the spino-parabrachio-amygdaloid and hypothalamic pathways in the autonomic and affective emotional aspects of pain. Prog Brain Res 107:243–255

Bernardi PS, Valtschanoff JG, Weinberg RJ, Schmidt HHHW, Rustioni A (1995) Synaptic interactions between primary afferent terminals and GABA and nitric oxide-synthesizing neurons in superficial laminae of the rat spinal cord. J Neurosci 15:1363–1371

Berthier M, Starkstein S, Leiguarda R (1988) Asymbolia for pain: a sensory-limbic disconnection syndrome. Ann Neurol 24:41–49

Besson JM (1999) The neurobiology of pain. Lancet 353:1610–1615

Besson JM, Chaouch A (1987) Peripheral and spinal mechanisms of nociception. Physiol Rev 67:67–186

Bester H, Besson JM, Bernard JF (1997a) Organization of efferent projections from the parabrachial area to the hypothalamus: a Phaseolus vulgaris-leucoagglutinin study in the rat. J Comp Neurol 383:245–281

Bester H, Matsumoto N, Besson JM, Bernard JF (1997b) Further evidence for the involvement of the spinoparabrachial pathway in nociceptive processes: a c-Fos study in the rat. J Comp Neurol 383:439–458

Bester H, Beggs S, Woolf CJ (2000) Changes in tactile stimuli-induced behavior and c-Fos expression in the superficial dorsal horn and in parabrachial nuclei after sciatic nerve crush. J Comp Neurol 428:45–61

Bester H, De Felipe C, Hunt SP (2001) The NK1 receptor is essential for the full expression of noxious inhibitory controls in the mouse. J Neurosci 21:1039–1046

Bevan S (1999) Nociceptive peripheral neurons: cellular properties. In: Wall PD, Melzack R (eds) Textbook of pain. Churchill-Livingstone, New York, pp 85–103

Birbaumer N, Lutzenberger W, Montoya P, Larbig W, Unerl K, Topfner S, Grodd W, Taub E, Flor H (1997) Effects of regional anesthesia on phantom limb pain are mirrored in changes in cortical reorganization. J Neurosci 17:5503–5508

Blakeman KH, Hao JX, Xu XJ, Jakoby AS, Shine J, Crawley JN, Iismaa T, Wiesenfeld-Hallin Z (2003) Hyperalgesia and increased neuropathic pain-like response in mice lacking galanin receptor 1 receptors. Neuroscience 117:221–227

Blomqvist A, Berkley KJ (1992) A re-examination of the spino-reticulo-diencephalic pathway in the cat. Brain Res 579:17–31

Blomqvist A, Craig AD (1991) Organization of spinal and trigeminal input to the PAG. In: Depaulis A, Bandler R (eds) The midbrain periaqueductal gray matter. Plenum Press, New York, pp 345–363

Blomqvist A, Craig AD (2000) Is neuropathic pain caused by the activation of nociceptive-specific neurons due to anatomic sprouting in the dorsal horn? J Comp Neurol 428:1–4

Blomqvist A, Ma W, Berkley KJ (1989) Spinal input to the parabrachial nucleus in the cat. Brain Res 480:29–36

Blomqvist A, Ericson AC, Craig AD, Broman J (1996) Evidence for glutamate as a neurotransmitter in spinothalamic tract terminals in the posterior region of owl monkeys. Exp Brain Res 108:33–44

Blomqvist A, Zhang ET, Craig AD (2000) Cytoarchitectonic and immunohistochemical characterization of a specific pain and temperature relay, the posterior portion of the ventral medial nucleus, in the human thalamus. Brain 123:601–619

Boddeke EW (2001) Involvement of chemokines in pain. Eur J Pharmacol 429:115–119

Bogousslavsky J, Regli F, Uske A (1988) Thalamic infarcts: clinical syndromes, etiology, and prognosis. Neurology 38:837–848

Boivie J (1978) Anatomical observations on the dorsal column nuclei. Their thalamic projection and the cytoarchitecture of some somatosensory thalamic nuclei in the monkey. J Comp Neurol 178:17–48

Boivie J (1979) An anatomic reinvestigation of the termination of the spinothalamic tract in the monkey. J Comp Neurol 168:343–370

Boivie J (1992) Hyperalgesia and allodynia in patients with CNS lesions. In: Willis WD (ed) Hyperalgesia and allodynia. Raven Press, New York, pp 363–373

Boivie J (1995) Pain syndromes in patients with CNS lesions and a comparison with nociceptive pain. In: Bromm B, Desmedt JE (eds) Pain and the brain: from nociception to cognition. Advances in pain research and therapy, vol 22. Raven Press, New York, pp 367–375

Boivie J (1999) Central pain. In: Wall PD, Melzack R (eds) Textbook of pain, 4th edn. Churchill Livingstone, Edinburgh, pp 879–914

Bonica JJ (1991) Semantic, epidemiologic and educational issues of central pain. In: Casey K (ed) Pain and central nervous system disease. The central pain syndromes. Raven Press, New York, pp 65–75

Bouhassira D, Attal N, Brasseur L, Parker F (2000) Quantitative sensory testing in patients with painful or painless syringomyelia. In: Devor M, Rowbotham MC, Wiesenfeld-Hallin Z (eds) Proceedings of the 9th world congress on pain. IASP Press, Seattle, pp 401–410

Boulton AJ, Ward JD (1986) Diabetic neuropathies and pain. Clin Endocrinol Metab 15:917–931

Bourgeais L, Gauriau C, Bernard JF (2001) Projections from the nociceptive area of the central nucleus of the amygdala to the forebrain: a PHA-L study in the rat. Eur J Neurosci 14:229–255.

Bourgeais L, Gauriau C, Monconduit L, Villanueva L, Bernard JF (2003) Dendritic domains of nociceptive-responsive parabrachial neurons match terminal fields of lamina I neurons in the rat. J Comp Neurol 464:238–256

Bowker RM, Westlund KN, Coulter JD (1981) Origin of serotonergic projections to the spinal cord in rat: an immunocytochemical-retrograde transport study. Brain Res 226:187–199

Bowker RM, Westlund KN, Sullivan MC, Wilbur JF, Coulter JD (1983) Descending serotonergic, peptidergic, and cholinergic pathways from the raphe nuclei: a multiple transmitter complex. Brain Res 288:33–48

Bowsher D (1957) Termination of the central pain pathway in man: the conscious appreciation of pain. Brain 80:606–622

Bowsher D (1978) Pain pathways and mechanisms. Anaesthesia 33:935–944

Bowsher D (1996) Central pain: clinical and physiological characteristics. J Neurol Neurosurg Psychiat 61:62–69

Bowsher D (1997) The management of postherpetic neuralgia. Postgrad Med J 73:623–629

Bowsher D (1999a) Central pain following spinal and supraspinal lesions. Spinal Cord 37:235–238

Bowsher D (1999b) Central post-stroke ("thalamic syndrome") and other central pains. Am J Hosp Palliat Care 16:593–597

Bowsher D (1999c) The lifetime occurrence of Herpes zoster and prevalence of post-herpetic neuralgia: a retrospective survey in an elderly population. Eur J Pain 3:335–342

Bowsher D, Leijon G, Thuomas KA (1998) Central poststroke pain. Correlation of MRI with clinical pain characteristics and sensory abnormalities. Neurology 51:1352–1358

Brodal A (1947) Central course of afferent fibers for pain in facialis, glossopharyngeal and vagus nerves. Clinical observations. Arch Neurol Psychiat 57:292–306

Brodal A (1981) Neurological anatomy in relation to clinical medicine, 3rd edn. Oxford University Press, New York

Broman J (1994) Neurotransmitters in subcortical somatosensory pathways. Anat Embryol 189:181–214

Broman J, Anderson S, Ottersen OP (1993) Enrichment of glutamate-like immunoreactivity in primary afferent terminals throughout the spinal cord dorsal horn. Eur J Neurosci 5:1050–1061

Bromm B, Lorenz J (1998) Neurophysiological evaluation of pain. Electroencephal Clin Neurophysiol 107:227–253

Bromm B, Scharein E, Vahle-Hinz C (2000) Cortex areas involved in the processing of normal and altered pain. Prog Brain Res 129:289–302

Brown AG (1981) Organization in the spinal cord. Springer, Berlin Heidelberg New York

Brown AG, Fyffe REW (1981) Form and function of dorsal horn neurones with axons ascending the dorsal columns in cat. J Physiol 321:31–47

Brown AG, Brown PB, Fyffe REW, Pubols LM (1983) Receptive field organization and response properties of spinal neurones with axons ascending the dorsal columns in the cat. J Physiol 377:575–588

Burstein R, Giesler GJ (1989) Retrogradely labeling of neurons in spinal cord that project directly to nucleus accumbens or the septal nuclei in the rat. Brain Res 497:149–154

Burstein R, Potrebic S (1993) Retrograde labeling of neurons in spinal cord that project directly to the amygdala or the orbital cortex in the rat. J Comp Neurol 335:469–485

Burstein R, Cliffer KD, Giesler GJ (1990a) Cells of origin of the spinohypothalamic tract in the rat. J Comp Neurol 291:329–344

Burstein R, Dado RJ, Giesler GJ (1990b) The cells of origin of the spinothalamic tract of the rat: a quantitative reexamination. Brain Res 511:329–337

Burstein R, Dado RJ, Cliffer KD, Giesler GJ (1991) Physiological characterization of spinohypothalamic tract neurons in the lumbar enlargement of rats. J Neurophysiol 66:261–284

Burstein R, Falkowsky O, Borsook D, Strassman A (1996) Distinct lateral and medial projections of the spinohypothalamic tract of the rat. J Comp Neurol 373:549–574

Bushnell MC, Duncan GH, Hofbauer RK, Ha B, Chen JI, Carrier B (1999) Pain perception: is there a role for primary somatosensory cortex? Proc Natl Acad Sci U S A 96:7705–7709

Byers MR (1984) Dental sensory receptors. Int Rev Neurobiol 25:39–94

Byers MR, Dong WK (1983) Autoradiographic location of sensory nerve endings in dentin of monkey teeth. Anat Rec 205:441–454

Cain DM, Wacnik PW, Turner M, Wendelschafer-Crabb G, Kennedy WR, Wilcox GL, Simone DA (2001) Functional interactions between tumor and peripheral nerve: changes in excitability and morphology of primary afferent fibers in a murine model of cancer pain. J Neurosci 21:9367–9376

Calcutt NA (2002) Potential mechanisms of neuropathic pain in diabetes. Int Rev Neurobiol 50:205–228

Cameron AA, Cliffer KD, Dougherty PM, Willis WD, Carlton SM (1991) Changes in lectin, GAP-43 and neuropeptide staining in the rat superficial dorsal horn following experimental peripheral neuropathy. Neurosci Lett 131:249–252

Cameron AA, Pover CM, Willis WD, Coggeshall RE (1992) Evidence that fine primary afferent axons innervate a wider territory in the superficial dorsal horn following peripheral axotomy. Brain Res 575:151–154

Campbell JN, Meyer RA (1996) Cutaneous nociceptors. In: Belmonte C, Cervero F (eds) Neurobiology of nociceptors. Oxford University Press, Oxford, pp 117–145

Cao YQ, Mantyh PW, Carlson EJ, Gillespie AM, Epstein CJ, Basbaum AI (1998) Primary afferent tachykinins are required to experience moderate to intense pain. Nature 392:334–335

Capra NF, Dessem D (1992) Central connections of trigeminal primary afferent neurons: topographical and functional considerations. Crit Rev Oral Biol Med 4:1–52

Carlstedt T, Cullheim S, Risling M (2004) Spinal cord in relation to the peripheral nervous system. In: Paxinos G, Mai JK (eds) The human nervous system, 2nd edn. Elsevier Academic Press, Amsterdam, pp 250–263

Carlton SM, Coggeshall RE (1999) Inflammation-induced changes in peripheral glutamate receptor populations. Brain Res 820:63–70

Carlton SM, Hargett GL, Coggeshall RE (2001) Localization of metabotropic glutamate receptors 2/3 on primary afferent axons in the rat. Neuroscience 105:957–969

Carstens E, Trevino DL (1978a) Laminar origins of spinothalamic projections in the cat as determined by the retrograde transport of horseradish peroxidase. J Comp Neurol 182:151–166

Carstens E, Trevino DL (1978b) Anatomical and physiological properties of ipsilaterally projecting spinothalamic neurons in the second cervical of the cat's spinal cord. J Comp Neurol 182:167–184

Casey KL (2000) Concepts of pain mechanisms: the contribution of functional imaging of the human brain. Prog Brain Res 129:277–287

Casey KL, Minoshima S, Berger KL, Koeppe RA, Morrow TJ, Frey A (1994) Positron emission tomographic analysis of cerebral structures activated specifically by repetitive noxious heat stimuli. J Neurophysiol 71:802–807

Casey KL, Minoshima S, Morrow TJ, Koeppe RA (1996) Comparison of human cerebral activation pattern during cutaneous warmth, heat pain, and deep cold pain. J Neurophysiol 76:571–581

Cassinari V, Pagni CA (1969) Central pain. A neurosurgical survey. Harvard University Press, Cambridge

Castro-Lopes JM, Coimbra A, Grant G, Arvidsson J (1990) Ultrastructural changes of the central scalloped (C1) primary afferent endings of synaptic glomeruli in the substantia gelatinosa Rolandi of the rat after peripheral neurotomy. J Neurocytol 19:329–337

Castro-Lopes JM, Tavares I, Coimbra A (1993) GABA decreases in the spinal cord dorsal horn after peripheral neurectomy. Brain Res 620:287–291

Castro-Lopes JM, Malcangio M, Pan BH, Bowery NG (1995) Complex changes of GABA A and GABA B receptor binding in the spinal cord dorsal horn following peripheral inflammation or neuroectomy. Brain Res 679:289–297

Catania MV, Tölle TR, Monyer H (1995) Differential expression of AMPA receptor subunit in NOS-positive neurons of cortex, striatum, and hippocampus. J Neurosci 15:7046–7061

Caterina MJ, Julius D (2001) The vanilloid receptor: a molecular gateway to the pain pathway. Annu Rev Neurosci 24:487–517

Caterina MJ, Schumacher MA, Tominaga M, Rosen TA, Levine JD, Julius D (1997) The capsaicin receptor: a heat activated ion channel in the pain pathway. Nature 389:816–824

Caterina MJ, Rosen TA, Tominaga M, Brake AJ, Julius D (1999) A capsaicin receptor homologue with a high threshold for noxious heat. Nature 398:436–441

Caterina MJ, Leffler A, Malmberg AB, Martin WJ, Trafton J, Petersen-Zeitz KR, Koltzenburg M, Basbaum AI, Julius D (2000) Impaired nociception and pain sensation in mice lacking the capsaicin receptor. Science 288:306–313

Cauna N (1973) The free penicillate nerve endings of the human hairy skin. J Anat 115:277–288

Cauna N (1980) Fine morphological characteristics and microtopography of the free nerve endings of the human digital skin. Anat Rec 198:643–656

Cervero F (1994) Sensory innervation of the viscera: peripheral basis of visceral pain. Physiol Rev 74:95–138

Cervero F (1996) Visceral nociceptors. In: Belmonte C, Cervero F (eds) Neurobiology of nociceptors. Oxford University Press, Oxford, pp 220–240

Cervero F, Jänig W (1992) Visceral nociceptors: a new world order? Trends Neurosci 15:374–378

Cesaro P, Mann MW, Moretii JL, Defer G, Roualdes B, Nguyen JP, Degos JD (1991) Central pain and thalamic hyperactivity: a single photon emission computerized tomographic study. Pain 47:329–336

Chabal C, Jacobson L, Russell LC, Burchiel KJ (1992) Pain response to perineuronal injection of normal saline, epinephrine, and lidocaine in humans. Pain 49:9–12

Chacur M, Milligan ED, Gazda LS, Armstrong C, Wang H, Tracey K, Maier SF, Watkins LR (2001) A new method of sciatic inflammatory neuritis (SIN): induction of unilateral and bilateral mechanical allodynia following acute unilateral peri-sciatic immune activation in rats. Pain 94:231–244

Chambers WW, Sprague JM (1955a) Functional localization in the cerebellum. I. Organization in longitudinal corticonuclear zones and their contribution to the control of posture, both extrapyramidal and pyramidal. J Comp Neurol 103:105–130

Chambers WW, Sprague JM (1955b) Functional localization in the cerebellum. II. Somatotopic organization in cortex and nuclei. Arch Neurol Psychiatry 74:653–680

Chaouch A, Menetrey D, Binder D, Besson JM (1983) Neurons at the origin of the medial component of the bulbopontine spinoreticular tract in the rat: an anatomical study using horseradish peroxidase retrograde transport. J Comp Neurol 214:309–320

Chen X, Levine JD (2001) Hyper-responsivity in a subset of C-fiber nociceptors in a model of painful diabetic neuropathy in the rat. Neuroscience 102:185–192

Cheunsuang O, Morris R (2000) Spinal lamina I neurons that express neurokinin 1 receptors: morphological analysis. Neuroscience 97:335–345

Cho HJ, Park EH, Bae MA, Kim JK (1996) Expression of mRNAs for preprotachykinin and nerve growth factor receptors in the dorsal root ganglion following peripheral inflammation. Brain Res 716:197–201

Cho HJ, Kim JK, Zhou XF, Rush RA (1997) Increased brain-derived neurotrophic factor immunoreactivity in rat dorsal root ganglia and spinal cord following peripheral inflammation. Brain Res 764:269–272

Choi B, Rowbotham MC (1997) Effect of adrenergic receptor activation on post-herpetic neuralgia pain and sensory disturbances. Pain 69:55–63

Choi Y, Raja SN, Moore LC, Tobin JR (1996) Neuropathic pain in rats is associated with altered nitric oxide synthase activity in neural tissue. J Neurol Sci 138:14–20

Chouchkov CN (1978) Cutaneous receptors. Adv Anat Embryol Cell Biol 54:3–61

Christensen MD, Hulsebosch CE (1997) Chronic central pain after spinal cord injury. J Neurotrauma 14:517–537

Chudler EH, Anton F, Dubner R, Kenshalo DR (1990) Responses of nociceptive SI neurons in monkeys and pain sensations in humans elicited by noxious thermal stimulation: effect of interstimulus interval. J Neurophysiol 63:559–569

Chung K, Coggeshall RE (1984) The ratio of dorsal root ganglion cells to dorsal root axons in sacral segments of the cat. J Comp Neurol 225:24–30

Chung K, Lee BH, Yoon YW, Chung JM (1996) Sympathetic sprouting in the dorsal root ganglia of the injured peripheral nerve in a rat neuropathic pain model. J Comp Neurol 376:241–252

Chung K, Yoon YW, Chung JM (1997) Sprouting sympathetic fibers form synaptic varicosities in the dorsal root ganglion of rats with neuropathic injury. Brain Res 751:275–280

Clark FM, Proudfit HK (1991) The projection of noradrenergic neurons in the A7 catecholamine cell group to the spinal cord in the rat demonstrated by anterograde tracing combined with immunocytochemistry. Brain Res 547:279–288

Clements JR, Magnusson KR, Hautman J, Beitz AJ (1991) Rat tooth pulp projections to spinal trigeminal subnucleus are glutamate-like immunoreactive. J Comp Neurol 309:281–288

Cliffer KD, Giesler GJ (1989) Postsynaptic dorsal column pathway of the rat. III. Distribution of ascending afferent fibers. J Neurosci 9:3146–3168

Cliffer KD, Willis WD (1994) Distribution of the postsynaptic dorsal column projection in the cuneate nucleus of monkeys. J Comp Neurol 345:84–93

Cliffer KD, Burstein R, Giesler GJ (1991) Distributions of spinothalamic, spinohypothalamic, and spinotelencephalic fibers revealed by anterograde transport of PHA-L in rats. J Neurosci 11:852–868

Clohisy DR, Mantyh PW (2003) Bone cancer pain. Cancer 97 [Suppl 3]:866–873

Coderre TJ, Katz J, Vaccarino AL, Melzack R (1993) Contribution of central neuroplasticity to pathological pain: review of clinical and experimental evidence. Pain 52:259–285

Coggeshall RE (1986) Nonclassical features of dorsal root ganglion cell organization. In: Yaksh TL (ed) Spinal afferent processing. Plenum, New York, pp 83–96

Coggeshall RE, Lekan HA, Doubell TP, Allchorne A, Woolf CJ (1997) Central changes in primary afferent fibers following peripheral nerve lesions. Neuroscience 77:1115–1122

Coggeshall RE, Lekan HA, White FA, Woolf CJ (2001) A-fiber sensory input induces neuronal cell death in the dorsal horn of the adult rat spinal cord. J Comp Neurol 435:276–282

Coghill RC, Talbot JD, Evans AC, Meyer E, Gjedde A, Bushnell MC, Duncan GH (1994) Distributed processing of pain and vibration by the human brain. J Neurosci 14:4095–4108

Coghill RC, Sang CN, Maisog JM, Iadarola MJ (1999) Pain intensity processing within the human brain: a bilateral, distributed mechanism. J Neurophysiol 82:1934–1943

Cohrs RJ, Randall J, Smith J, Gilden DH, Dabrowski C, van der Keyl H, Tal-Singer R (2000) Analysis of individual human trigeminal ganglia for latent herpes simplex virus type 1 and varicella-zoster virus nucleic acids using real-time PCR. J Virol 74:11464–11471

Coimbra A, Sodre-Borges BP, Magelhaes MM (1974) The substantia gelatinosa Rolandi of the rat. Fine structure, cytochemistry (acid phosphatase) and changes after dorsal root section. J Neurocytol 3:199–217

Colburn RW, Rickman AJ, DeLeo JA (1999) The effect of site and type of nerve injury on spinal glial activation and neuropathic pain behavior. Exp Neurol 157:289–304

Colvin LA, Mark MA, Duggan AW (1996) Bilaterally enhanced dorsal horn postsynaptic currents in a rat model of peripheral mononeuropathy. Neurosci Lett 207:29–32

Conn PJ, Pin JP (1997) Pharmacology and functions of metabotropic glutamate receptors. Annu Rev Pharmacol Toxicol 37:205–237

Conti F, De Biasi S, Giuffrida R, Rustioni A (1990) Substance P-containing projections in the dorsal columns of rats and cats. Neuroscience 34:607–621

Coppes MH, Marani E, Thomeer TR, Oudega M, Groen GJ (1990) Innervation of annulus fibrosus in low back pain. Lancet 336:189–190

Coppes MH, Marani E, Thomeer TR, Groen GJ (1997) Innervation of "painful" lumbar discs. Spine 22:2342–2349

Craig AD (1987) Medial thalamus and nociception: the nucleus submedius. In Besson JM, Guilbaud G, Peschanski M (eds) Thalamus and pain. Elsevier, Amsterdam, pp 227–243

Craig AD (1991) Spinal distribution of ascending lamina I axons anterogradely labeled with Phaseolus vulgaris leucoagglutinin (PHA-L) in the cat. J Comp Neurol 313:377–393

Craig AD (1992) Spinal and trigeminal lamina I input to the locus coeruleus anterogradely labeled with Phaseolus vulgaris leucoagglutinin (PHA-L) in the cat and the monkey. Brain Res 584:325–328

Craig AD (1995) Distribution of brainstem projections from spinal lamina I neurons in the cat and monkey. J Comp Neurol 361:225–248

Craig AD (1996a) Pain, temperature, and the sense of the body. In: Franzen O, Johansson R, Terenius L (eds) Somesthesis and the neurobiology of the somatosensory cortex. Birkhäuser, Basel, pp 27–39

Craig AD (1996b) An ascending general homeostatic afferent pathway originating in lamina I. Prog Brain Res 107:225–242

Craig AD (1998) A new version of the thalamic disinhibition hypothesis of central pain. Pain Forum 7:1–14

Craig AD (2000) The functional anatomy of lamina I and its role in post-stroke central pain. Prog Brain Res 129:137–151

Craig AD (2003a) Pain mechanisms: labeled lines versus convergence in central processing. Annu Rev Neurosci 26:1–30

Craig AD (2003b) Distribution of trigeminothalamic and spinothalamic lamina I terminations in the cat. Somatosens Mot Res 20:209–222

Craig AD (2003c) Interoception: the sense of the physiological condition of the body. Curr Opin Neurobiol 13:500–505

Craig AD (2003d) A new view of pain as a homeostatic emotion. Trends Neurosci 26:303–307

Craig AD (2004) Lamina I, but not lamina V, spinothalamic neurons exhibit responses that correspond with burning pain. J Neurophysiol 92:2604–2609

Craig AD, Dostrovsky JO (1999) Medulla to thalamus. In Wall PD, Melzack R (eds) Textbook of pain. Churchill Livingstone, Edinburgh, pp 183–214

Craig AD, Bushnell MC, Zhang ET, Blomqvist A (1994) A thalamic nucleus specific for pain and temperature sensation. Nature 372:770–773

Craig AM, Blackstone CD, Huganir RL, Banker G (1993) The distribution of glutamate receptors in cultured rat hippocampal neurons: postsynaptic clustering of AMPA-selective subunits. Neuron 10:1055–1068

Craig AM, Blackstone CD, Huganir RL, Banker G (1994) Selective clustering of glutamate and g-aminobutyric acid receptors opposite terminals releasing the corresponding neurotransmitters. Proc Natl Acad Sci U S A 91:12373–12377

Croen KD, Ostrove JM, Dragovic LJ, Straus SE (1988) Patterns of gene expression and sites of latency in human nerve ganglia are different for varicella-zoster and herpes simplex viruses. Proc Natl Acad Sci U S A 85:9773–9777

Croul S, Sverstiuk A, Radzievsky A, Murray M (1995) Modulation of neurotransmitter receptors following unilateral L1-S2 deafferentation: NK1, NK3, NMDA, and 5HT1a receptor binding autoradiography. J Comp Neurol 361:633–644

Csillik B, Janka Z, Boncz I, Kalman J, Mihally A, Vecsei A, Knyihar E (2003) Molecular plasticity of primary nociceptive neurons: relations of the NGF-c-jun system to neurotomy and chronic pain. Ann Anat 185:303–314

Cui M, Feng Y, McAdoo D, Willis W (1999) Periaqueductal gray stimulation-induced inhibition of nociceptive dorsal horn neurons in rats is associated with the release of norepinephrine, serotonin and amino acids. J Pharmacol Exp Ther 289:868–876

Dado RJ, Giesler GJ (1990) Afferent input to nucleus submedius in rats: retrograde labeling of neurons in the spinal cord and caudal medulla. J Neurosci 10:2672–2686

Dado RJ, Katter JT, Giesler GJ (1994a) Spinothalamic and spinohypothalamic tract neurons in the cervical enlargement of rats. I. Locations of antidromically identified axons in the thalamus and hypothalamus. J Neurophysiol 71:959–980

Dado RJ, Katter JT, Giesler GJ (1994b) Spinothalamic and spinohypothalamic tract neurons in the cervical enlargement of rats. II. Responses to innocuous and noxious mechanical and thermal stimuli. J Neurophysiol 71:981–1002

Dado RJ, Katter JT, Giesler GJ (1994c) Spinothalamic and spinohypothalamic tract neurons in the cervical enlargement of rats. III. Locations of antidromically identified axons in the cervical cord white matter. J Neurophysiol 71:1003–1021

Dahlström A, Fuxe K (1964) Evidence for the existence of monoamine-containing neurons in the mammalian nervous system. I. Demonstration of monoamines in the cell bodies of brain stem neurons. Acta Physiol Scand 232 (Suppl):1–55

Dallel R, Raboisson P, Auroy P, Woda A (1988) The rostral part of the trigeminal sensory complex is involved in orofacial nociception. Brain Res 448:7–19

Dalsgaard CJ, Jernbeck J, Stains W, Kjartansson J, Haegerstrand A, Hökfelt T, Brodin E, Cuello AC, Brown JC (1989) Calcitonin gene-related peptide-like immunoreactivity in nerve fibers in the human skin. Relation to fibers containing substance P-, somatostatin- and vasoactive intestinal polypeptide-like immunoreactivity. Histochemistry 91:35–38

Darian-Smith C (2004) Primary afferent terminal sprouting after a cervical dorsal rootlet section in the macaque monkey. J Comp Neurol 470:134–150

Darian-Smith C, Brown S (2000) Functional changes at periphery and cortex following dorsal root lesions in adult monkeys. Nat Neurosci 3:476–481

Davis KD (2000) The neural circuitry of pain as explored with functional MRI. Neurol Res 22:313–317

Davis KD, Meyer RA, Campbell JN (1993) Chemosensitivity and sensitization of nociceptive afferents that innervate the hairy skin of monkey. J Neurophysiol 69:1071–1081

De Biasi S, Rustioni A (1988) Glutamate and substance P coexist in primary afferent terminals in the superficial laminae of the spinal cord. Proc Natl Acad Sci U S A 85:7820–7824

De Biasi S, Amadeo A, Spreafico R, Rustioni A (1994) Enrichment of glutamate immunoreactivity in lemniscal terminals in the ventropostero lateral thalamic nucleus of the rat: an immunogold and WGA-HRP study. Anat Rec 240:131–140

Decosterd I, Woolf CJ (2000) Spared nerve injury: an animal model of persistent peripheral neuropathic pain. Pain 87:149–158

Decosterd I, Allchorne A, Woolf CJ (2002) Progressive tactile hypersensitivity after a peripheral nerve crush: non-noxious mechanical stimulus-induced neuropathic pain. Pain 100:155–162

Defrin R, Ohry A, Blumen N, Urca G (2001) Characterization of chronic pain and somatosensory function in spinal cord injury subjects. Pain 89:253–263

DeJong RN (1977) CNS manifestations of diabetes mellitus. Postgrad Med 61:101–107

De La Blanchardiere A, Rozenberg F, Caumes E, Picard O, Lionnet F, Livartowski J, Coste J, Sicard D, Lebon P, Salmon-Ceron D (2000) Neurological complications of varicella-zoster virus infection in adults with human immunodeficiency virus infection. Scand J Infect Dis 32:263–269

Dejerine PJ (1914) Sémiologie des affections du système nerveux. Masson, Paris

Dejerine J, Roussy J (1906) Le syndrome thalamique. Rev Neurol 14:521–532

DeLeo JA, Colburn RW (1999) Proinflammatory cytokines and glial cells: their role in neuropathic pain. In: Watkins L (ed) Cytokines and pain. Birkhauser, Basel, pp 159–182

Derbyshire SWG, Jones AKP (1998) Cerebral responses to a continual tonic pain stimulus measured using positron emission tomography. Pain 76:127–135

Devor M (1994) The pathophysiology of damaged peripheral nerves. In: Wall PD, Melzack R (eds) Textbook of pain, 3rd edn. Churchill Livingstone, Edinburgh, pp 79–100

Devulder JE (2002) Postherpetic ophthalmic neuralgia. Bull Soc Belge Ophtalmol 285:19–23

DiFiglia M, Aronin N (1990) Synaptic interactions between GABAergic neurons and trigeminothalamic cells in the rat trigeminal nucleus caudalis. Synapse 6:358–363

Ding YQ, Nomura S, Kaneko T, Mizuno N (1995a). Co-localization of mu-opioid receptor-like and substance P-like immunoreactivities in axon terminals within the superficial layers of the medullary and spinal dorsal horns of the rat. Neurosci Lett 198:45–48

Ding YQ, Takada M, Shigemot R, Mizuno N (1995b) Spinoparabrachial tract neurons showing substance P receptor-like immunoreactivity in the lumbar spinal cord of the rat. Brain Res 674:336–340

Donnerer J, Stein C (1992) Evidence for an increase in the release of CGRP from sensory nerves during inflammation. Ann N Y Acad Sci 657:505–506

Donnerer J, Schuligoi R, Stein C, Amann R (1993) Upregulation, release and axonal transport of substance P and calcitonin gene-related peptide in adjuvant inflammation and regulatory function of nerve growth factor. Regul Pept 46:150–154

Dohrn CS, Mullet MA, Price RH, Beitz AJ (1994) Distribution of nitric oxide synthase-immunoreactive interneurons in the spinal trigeminal nucleus. J Comp Neurol 346:449–460

Dostrovsky JO (2000) Role of thalamus in pain. Prog Brain Res 129:245–257

Doubell TP, Mannion RJ, Woolf CJ (1997) Intact sciatic myelinated primary afferent terminals collaterally sprout in the adult rat dorsal horn following section of a neighbouring peripheral nerve. J Comp Neurol 380:95–104

Doubell TP, Mannion RJ, Woolf CJ (1999) The dorsal horn: state dependent sensory processing, plasticity and the generation of pain. In: Wall PD, Melzack R (eds) The textbook of pain. Churchill Livingstone, London, pp 165–182

Dougherty PM, Schwartz A, Lenz FA (1999) Responses of primate spinomesencephalic tract cells to intradermal capsaicin. Neuroscience 90:1377–1392

Dray A (1995) Inflammatory mediators of pain. Br J Anaesth 75:125–131

Druschky K, Lang E, Hummel C, Kaltenhäuser M, Kohllöffel LUE, Neundörfer B, Stefan H (2000) Pain-related somatosensory evoked magnetic fields induced by controlled ballistic mechanical impacts. J Clin Neurophysiol 17:613–622

Duce I, Keen P (1977) An ultrastructural classification of the neuronal cell bodies of the rat dorsal root ganglion using zinc iodide-osmium impregnation. Cell Tissue Res 185:263–277

Dueland AN, Ranneberg-Nilsen T, Degre M (1995) Detection of latent varicella zoster virus DNA and human gene sequences in human trigeminal ganglia by in situ amplification combined with in situ hybridization. Arch Virol 140:2055–2066

Dun NJ, Dun SL, Chiba T, Forstermann U (1995) Nitric oxide synthase-immunoreactive vagal afferent fibers in rat superior cervical ganglia. Neuroscience 65:231–239

Dworkin RH (2002) An overview of neuropathic pain: syndromes, symptoms, signs and several mechanisms. Clin J Pain 18:343–349

Dworkin RH, Johnson RW (1999) A belt of roses from hell: pain in herpes zoster and postherpetic neuralgia. In: Block AR, Kremer EF, Fernandez E (eds) Handbook of pain syndromes: biopsychosocial perspectives. Erlbaum, Hillsdale, New Jersey, pp 371–402

Dworkin RH, Portenoy RK (1996) Pain and its persistence in herpes zoster. Pain 67:241–251

Dworkin RH, Schmader KE (2003) Treatment and prevention of postherpetic neuralgia. Clin Infect Dis 36:877–882

Dworkin RH, Carrington D, Cunningham A, Kost RG, Levin MJ, McKendrick MW, Oxman MN, Rentier B, Schmader KE, Tappeiner G, Wassilew SW, Whitley RJ (1997) Assessment of pain in herpes zoster: lessons learned from antiviral trials. Antiviral Res 33:73–85

Dworkin RH, Perkins FM, Nagasako EM (2000) Prospects for the prevention of postherpetic neuralgia in herpes zoster patients. Clin J Pain 16 (Suppl):S90-S100

Eaton SE, Harris ND, Rajbhandai SM, Greenwood P, Wilkinson ID, Ward JD, Griffiths PD, Tesfaye S (2001) Spinal cord involvement in diabetic peripheral neuropathy. Lancet 358:35–36

Eckersley L (2002) Role of the Schwann cell in diabetic neuropathy. Int Rev Neurobiol 50:293–321

Edgar RE, Best LG, Quail PA, Obert AD (1993) Computer-assisted DREZ microcoagulation: posttraumatic spinal deafferentation pain. J Spin Dis 6:48–56

Edvinsson L, Mulder H, Goadsby PJ, Uddman R (1998) Calcitonin gene-related peptide and nitric oxide in the trigeminal ganglion: cerebral vasodilatation from trigeminal nerve stimulation involves mainly calcitonin gene-related peptide. J Auton Nerv Syst 70:15–22

Eide PK, Jorum E, Stenehjem AE (1996) Somatosensory findings in patients with spinal cord injury and central dysaesthesia pain. J Neurol Neurosurg Psychiat 60:411–415

Ekerot CF, Gustavsson P, Oscarsson O, Schouenborg J (1987a) Climbing fibres projecting to cat cerebellar anterior lobe activated by cutaneous A and C fibres. J Physiol 368:529–538

Ekerot CF, Oscarsson O, Schouenborg J (1987b) Stimulation of cat cutaneous nociceptive fibres causing tonic and synchronous activity in climbing fibres. J Physiol 386:539–546

Ekerot CF, Garwicz M, Schouenborg J (1991) The postsynaptic dorsal column pathway mediates cutaneous nociceptive information to cerebellar climbing fibres in the cat. J Physiol 441:275–284

Engelman HS, Allen TB, MacDermott AB (1999) The distribution of neurons expressing calcium-permeable AMPA receptors in the superficial laminae of the spinal cord dorsal horn. J Neurosci 19:2081–2089

Ericson AC, Blomqvist A, Craig AD, Ottersen OP, Broman J (1995) Evidence for glutamate as transmitter in trigemino- and spinothalamic tract terminals in the nucleus submedius of cats. Eur J Neurosci 7:305–317

Ericson AC, Blomqvist A, Krout K, Craig AD (1996) Fine structural organization of spinothalamic and trigeminothalamic lamina I terminations in the nucleus submedius of the cat. J Comp Neurol 371:497–512

Eriksson NP, Persson JK, Svensson M, Arvidsson J, Molander C, Aldskogius H (1993) A quantitative analysis of the microglial cell reaction in central primary sensory projection territories following peripheral nerve injury in the adult rat. Exp Brain Res 96:19–27

Esiri MM, Kennedy PGE (1997) Viral diseases. In: Graham DI, Lantos PL (eds) Greenfield's neuropathology, 6th edn, vol 2. Arnold, London, pp 3–63

Esiri MM, Tomlinson AH (1972) Herpes zoster. Demonstration of virus in trigeminal nerve and ganglion by immunofluorescence and electron microscopy. J Neurol Sci 15:35–48

Fabian VA, Wood B, Crowley P, Kakulas BA (1997) Herpes zoster brachial plexus neuritis. Clin Neuropathol 16:61–64

Farel PB (2002) Trust, but verify: the necessity of empirical verification in quantitative neurobiology. Anat Rec 269:157–161

Farrar JT, Portenoy RK (2001) Neuropathic cancer pain: the role of adjuvant analgesics. Oncology (Huntingt) 15:1435–1442

Feil K, Herbert H (1995) Topographic organization of spinal and trigeminal somatosensory pathways to the rat parabrachial and Kölliker-Fuse nuclei. J Comp Neurol 353:506–528

Feirabend HKP, Marani E (2003) Dorsal root ganglia. In: Aminoff M, Daroff R (eds) Encyclopedia of the neurological sciences, vol 2. Academic Press, San Diego, pp 28–33

Feirabend HKP, Kok P, Choufoer H, Ploeger S (1994) Preservation of myelinated fibers for electron microscopy: a qualitative comparison of aldehyde fixation, microwave stabilization and other procedures all complete by osmification. J Neurosci Methods 55:137–153

Feirabend HKP, Choufoer H, Ploeger S (1994) Preservation and staining of myelinated fibers. Methods 15:123–131

Feldman EL, Russell JW, Sullivan KA, Golovoy D (1999) New insights into the pathogenesis of diabetic neuropathy. Curr Opin Neurol 12:553–563

Fields HL, Rowbotham M, Baron R (1998) Postherpetic neuralgia: irritable nociceptors and deafferentation. Neurobiol Dis 5:209–227

Fields HL (1992) Is there a facilitating component in central pain modulation? Am Pain Soc J 1:71–78

Fields HL (2000) Pain modulation: expectation, opioid analgesia and virtual pain. Prog Brain Res 122:245–253

Fileds HL, Basbaum AI (1999) Central nervous system mechanisms of pain modulation. In: Wall PD, Melzack R (eds) Textbook of pain. Churchill-Livingstone, New York, pp 309–329

Finnerup NB, Johannesen IL, Sindrup SH, Bach FW, Jensen TS (2001) Pain and dysesthesia in patients with spinal cord injury: a postal survey. Spinal Cord 39:256–262

Finnerup NB, Johannesen IL, Fuglsang-Frederiksen A, Bach FW, Jensen TS (2003) Sensory function in spinal cord injury patients with and without central pain. Brain 126:57–70

Finnerup NB, Jensen TS (2004) Spinal cord injury pain – mechanisms and treatment. Eur J Neurol 11:73–82

Fitzgerald M, Woolf CJ, Shortland P (1990) Collateral sprouting of the central terminals of cutaneous primary afferent neurons in the rat spinal cord: pattern, morphology, and influence of targets. J Comp Neurol 300:370–385

Flor H (2003) Cortical reorganization and chronic pain: implications for rehabilitation. J Rehabil Med S41:66–72

Flor H, Elbert T, Knecht S, Wienbruch C, Pantev C, Birbaumer N, Larbig W, Taub E (1995) Phantom limb pain as a perceptual correlate of cortical reorganization following arm amputation. Nature 375:482–484

Flor H, Braun C, Elbert T, Birbaumer N (1997) Extensive reorganization of primary sensory cortex in chronic back pain patients. Neurosci Lett 224:5–8

Florence SL, Garraghty PE, Carlson M, Kaas JH (1993) Sprouting of peripheral nerve axons in the spinal cord of monkeys. Brain Res 601:343–348

Florence SL, Kaas JH (1995) Large-scale reorganization at multiple levels of the somatosensory pathway follows therapeutic amputation of the hand in monkeys. J Neurosci 15:8083–8095

Foerster O (1936) Motorische Felder und Bahnen. In: Bumke O, Foerster O (eds) Handbuch der Neurologie, vol 6. Springer, Berlin, pp 1–357

Foster GA, Sizer AR, Rees H, Roberts MHT (1989) Afferent projections to the rostral anterior pretectal nucleus of the rat: a possible role in the processing of noxious stimuli. Neuroscience 29:685–694

Fox A, Eastwood C, Gentry C, Manning D, Urban L (1999) Critical evaluation of the streptozotocin model of painful diabetic neuropathy in the rat. Pain 81:307–316

Fraher JP (1992) The CNS-PNS transitional zone of the rat. Morphometric studies at cranial and spinal level. Prog Neurobiol 38:261–316

Fraher JP (2000) The transitional zone and CNS regeneration. J Anat 196:137–158

Freemont AJ, Peacock TE, Goupille P, Hoyland JA, O'Brien J, Jayson MI (1997) Nerve ingrowth into diseased intervertebral disc in chronic back pain. Lancet 350:178–181

Fricke B, Andres KH, Von During M (2001) Nerve fibers innervating the cranial and spinal meninges: morphology of nerve fiber terminals and their structural integration. Microsc Res Tech 53:96–105

Fu K-Y, Light AR, Matsushima GK, Maixner W (1999) Microglial reactions after subcutaneous formalin injection into the rat hindpaw. Brain Res 825:59–67

Fulwiler CE, Saper CB (1984) Subnuclear organization of the efferent connections of the parabrachial nucleus in the rat. Brain Res Rev 7:229–259

Furuyama T, Kiyama H, Sato K, Park HT, Maeno H, Takagi H, Tohyama M (1993) Region-specific expression of subunits of ionotropic glutamate receptors (AMPA-type, KA-type and NMDA receptors) in the rat spinal cord with special reference to nociception. Mol Brain Res 18:141–151

Garcin R (1968) Thalamic syndrome and pain of central origin. In: Soulairac A, Cahn J, Charpentier J (eds) Pain. Academic Press, London, pp 521–541

Gardell LR, Vanderah TW, Gardell SE, Wang R, Ossipov MH, Lai J, Porreca F (2003) Enhanced evoked excitatory transmitter release in experimental neuropathy requires descending facilitation. J Neurosci 23:8370–8379

Garrett L, Coggeshall RE, Patterson JT, Chung K (1992) Numbers and proportions of unmyelinated axons at cervical levels in the funiculus gracilis of monkey and cat. Anat Rec 232:301–304

Garrison CJ, Dougherty PM, Kaander KC, Carlton SM (1991) Staining of glial fibrillary acidic protein (GFAP) in lumbar spinal cord increases following a sciatic nerve constriction injury. Brain Res 565:1–7

Garrison CJ, Dougherty PM, Carlton SM (1993) Quantitative analysis of substance P and calcitonin gene-related peptide immunohistochemical staining in the dorsal horn of neuropathic MK-801-treated rats. Brain Res 607:205–214

Gauriau C, Bernard JF (2002) Pain pathways and parabrachial circuits in the rat. Exp Physiol 87:251–258

Gauriau C, Bernard JF (2004) A comparative reappraisal of projections from the superficial laminae of the dorsal horn in the rat: the forebrain. J Comp Neurol 468:24–56

Gebhart GF (1996) Visceral polymodal receptors. Prog Brain Res 113:101–112

Gerhart KD, Yezierski RP, Fang ZR, Willis WD (1983) Inhibition of primate spinothalamic tract neurons by stimulation in ventral posterior lateral (VPLc) thalamic nucleus: possible mechanisms. J Neurophysiol 49:406–423

Ghatak NR, Zimmerman HM (1973) Spinal ganglion in herpes zoster. Arch Pathol 95:411–415

Giamberardino MA, Vecchiet L (1996) Pathophysiology of visceral pain. Curr Rev Pain 1:23–33

Gibbins IL, Wattchow D, Coventry B (1987) Two immunohistochemically identified populations of calcitonin gene-related peptide (CGRP)-immunoreactive axons in human skin. Brain Res 414:143–148

Giesler GJ, Cliffer KD (1985) Postsynaptic dorsal column pathway of the rat. II. Evidence against an important role in nociception. Brain Res 326:347–356

Giesler GJ Jr, Menetrey D, Basbaum AI (1979) Differential origins of spinothalamic tract projections to medial and lateral thalamus in the rat. J Comp Neurol 184:107–126

Giesler GJ, Yezierski RP, Gerhart KD, Willis WD (1981) Spinothalamic tract neurons that project to medial and/or lateral thalamic nuclei: evidence for a physiologically novel population of spinal cord neurons. J Neurophysiol 46:1285–1308

Giesler GJ, Nahin RL, Madsen AM (1984) Postsynaptic dorsal column pathway of the rat. I. Anatomical studies. J Neurophysiol 51:260–275

Gilden DH, Vafai A, Shtram Y, Becker Y, Devlin M, Wellish M (1983) Varicella-zoster virus DNA in human sensory ganglia. Nature 306:478–480

Gilden DH, Rozenman Y, Murray R, Devlin M, Vafai A (1987) Detection of varicella-zoster virus nucleic acid in neurons of normal human thoracic ganglia. Ann Neurol 22:377–380

Gilden DH, Dueland AN, Devlin ME, Mahalingham R, Cohrs RJ (1992) Varicella-zoster virus reactivation without rash. J Infect Dis 166 (Suppl 1):S30-S34

Gilden DH, Kleinschmidt-DeMasters BK, Wellish M, Hedley-Whyte ET, Rentier B, Mahalingham R (1996) Varicella zoster virus, a cause of waxing and waning vasculitis: the N England Journal of Medicine case 5–1995 revisited. Neurology 47:1441–1446

Gilden DH, Kleinschmidt-DeMasters BK, LaGuardia JJ, Mahalingham R, Cohrs RJ (2000) Neurologic complications of the reactivation of varicella-zoster virus. N Engl J Med 342:635–645

Gilden DH, Cohrs RJ, Hayward AR, Wellish M, Mahalingam R (2003) Chronic varicella-zoster virus ganglionitis – a possible cause of postherpetic neuralgia. J Neurovirol 9:404–407

Giuffrida R, Rustioni A (1992) Dorsal root ganglion neurons projecting to the dorsal column nuclei. J Comp Neurol 316:206–220

Glees P, Bailey RA (1951) Schichtung und Fasergröße des Tractus spino-thalamicus des Menschen. Mschr Psychiat 122:129–140

Gobel S (1978a) Golgi studies of the neurons in layer I of the dorsal horn of the medulla (trigeminal nucleus caudalis). J Comp Neurol 180:375–394

Gobel S (1978b) Golgi studies of the neurons in layer II of the dorsal horn of the medulla (trigeminal nucleus caudalis). J Comp Neurol 180:395–414

Gobel S, Falls WM, Hockfield S (1977) The division of the dorsal and ventral horns of the mammalian caudal medulla into eight layers using anatomical criteria. In: Anderson DJ, Matthews B (eds) Pain in the trigeminal region. Elsevier, Amsterdam, pp 443–453

Goldschneider A (1881) Zur Lehre von den spezifischen Energien der Sinnesorgane. Schumacher, Berlin (quoted from Bowsher, 1978)

Gonzalo-Riuz A, Lieberman AR (1995) GABAergic projections from the thalamic reticular nucleus to the anteroventral and anterodorsal thalamic nuclei of the rat. J Chem Neuroanat 9:165–174

Gordh T, Sharma HS, Alm P, Westman J (1998) Spinal nerve lesion induces upregulation of neuronal nitric oxide synthase in the spinal cord. An immunohistochemical investigation in the rat. Amino Acids 14:105–112

Gordh T, Sharma HS, Azizi M, Alm P, Westman J (2000) Spinal nerve lesion induces upregulation of constitutive isoform of heme oxygenase in the spinal cord. An immunohistochemical investigation in the rat. Amino Acids 19:373–381

Grachev ID, Fredrickson BE, Apkarian AV (2000) Abnormal brain chemistry in chronic back pain: an in vivo proton magnetic resonance spectroscopy study. Pain 89:7–18

Gradl G, Gaida S, Gierer P, Mittlmeier T, Vollmar B (2004) In vivo evidence for apoptosis, but no inflammation in the hindlimb muscles of neuropathic rats. Pain 112:121–130

Granum S (1986) The spinothalamic system of the rat. I. Locations of cells of origin. J Comp Neurol 247:159–180

Graziano A, Jones EG (2004) Widespread thalamic terminations of fibers arising in the superficial medullary dorsal horn of monkeys and their relation to calbindin immunoreactivity. J Neurosci 24:248–256

Greenspan JD, Lee RR, Lenz FA (1999) Pain sensitivity alterations as a function of lesion location in the parasylvian cortex. Pain 81:273–282

Greiff (1883) Zur Lokalization der Hemichorea. Arch Psych Nervenkr 14:598 (quoted from Boivie, 1999)

Griffiths G (1993) Fine structure immunocytochemistry. Springer, Berlin

Guillery RW, Harting (2003) Structure and connections of the thalamic reticular nucleus: advancing views over half a century. J Comp Neurol 463:360–371

Guillery RW, Feig SL, Lozsadi DA (1998) Paying attention to the thalamic reticular nucleus. Trends Neurosci 21:28–32

Guo A, Vulchanova L, Wang J, Li X, Elde R (1999) Immunocytochemical localization of vanilloid receptor 1 (VR1): relationship to neuropeptides, the P2X3 purinoceptor and IB4 binding sites. Eur J Neurosci 11:946–958

Guthoff R, Wienss H, Hahnel C, Wree A (2005) Epithelial innervation of human cornea: a three-dimensional study using confocal laser scanning fluorescence microscopy. Cornea 24:608–613

Hains BC, Willis WD, Hulsebosch CE (2003a) Serotonin receptors 5-HT1A and 5HT3 reduce hyperexcitability of dorsal horn neurons after chronic spinal cord hemisection injury in rat. Exp Brain Res 149:174–186

Hains BC, Willis WD, Hulsebosch CE (2003b) Temporal plasticity of dorsal horn somatosensory neurons after acute and chronic spinal cord hemisection in rat. Brain Res 970:238–241

Halata Z (1975) The Mechanoreceptors of the mammalian skin. Ultrastructure and morphological classification. Adv Anat Embryol Cell Biol 50:1–77

Halata Z, Munger BL (1986) The neuroanatomical basis for the protopathic sensibility of the human glans penis. Brain Res 371:205–230

Halata Z, Wagner C, Baumann KI (1999) Sensory nerve endings in the anterior cruciate ligament (Lig. cruciatum anterius) of sheep. Anat Rec 254:13–21

Halata Z, Grim M, Bauman KI (2003) Friedrich Sigmund Merkel and his "Merkel cell", morphology, development, and physiology: review and new results. Anat Rec 271A:225–239

Han ZS, Zhang ET, Craig AD (1998) Nociceptive and thermoreceptive lamina I neurons are anatomically distinct. Nature Neurosci 1:218–225

Harden RN (2005) Chronic neuropathic pain. Mechanisms, diagnosis, and treatment. Neurologist 11:111–122

Haring JH, Henderson TA, Jacquin MF (1990) Principalis- or parabrachial-projecting spinal trigeminal neurons do not stain for GABA or GAD. Somatosens Mot Res 7:391–397

Harper AA, Lawson SN (1985) Conduction velocity is related to morphological cell type in rat dorsal root ganglion neurones. J Physiol 359:31–46

Harris JA, Corsi M, Quartaroll M, Arban R, Bentivoglio M (1996) Upregulation of spinal glutamate receptors in chronic pain. Neuroscience 74:7–12

Hashizume H, DeLeo JA, Colburn RW, Weinstein JN (2000) Spinal glial activation and cytokine expression after lumbar root injury in the rat. Spine 25:1206–1217

Hassler R (1959) Anatomy of the thalamus. In: Schaltenbrand G, Bailey (eds) Introduction to stereotaxic operations with an atlas of the human brain. Thieme, Stuttgart, pp 230–290

Hassler R (1960) Die zentralen Systeme des Schmerzes. Acta Neurochir 8:354–423

Hassler R (1982) Cytoarchitectonic organization of the thalamic nuclei. In: Schaltenbrand G, Walker AE (eds) Stereotaxy of the human brain. Thieme, Stuttgart, pp 140–180

Head H, Campbell AW (1900) The pathology of herpes zoster and its bearing on sensory localization. Brain 23:353–523 (quoted from Oaklander, 1999)

Head H, Holmes G (1911) Sensory disturbances from cerebral lesions. Brain 34:102–254

Helliwell RJA, McLatchie LM, Clarke M, Winter J, Bevan S, McIntyre P (1998) Capsaicin sensitivity is associated with the expression of the vanilloid (capsaicin) receptor (VR1) mRNA in adult rat sensory ganglia. Neurosci Lett 250:177–180

Henley JM, Jenkins R, Hunt SP (1993) Localization of glutamate receptor binding sites and mRNAs to the dorsal horn of the rat spinal cord. Neuropharmacology 32:37–41

Heppelmann B, Messlinger K, Neiss WF, Schmidt RF (1990) Ultrastructural three-dimensional reconstruction of group III and group IV sensory nerve endings ("free nerve endings") in the knee joint capsule of the cat: evidence for multiple receptive sites. J Comp Neurol 292:103–116

Heppelmann B, Messlinger K, Neiss WF, Schmidt RF (1994) Mitochondria in fine afferent nerve fibres of the knee joint in the cat: a quantitative electron-microscopical examination. Cell Tissue Res 275:493–501

Heppelmann B, Messlinger K, Neiss WF, Schmidt RF (1995) Fine sensory innervation of the knee joint capsule by group III and IV nerve fibers in the cat. J Comp Neurol 351:415–428

Hestrin S (1992) Activation and desensitization of glutamate-activated channels mediating fast excitatory synaptic currents in the visual cortex. Neuron 9:991–999

Hill EL, Elde R (1991) Distribution of CGRP-, VIP-, D beta H-, SP-, and NPY-immunoreactive nerves in the periosteum of the rat. Cell Tissue Res 264:469–480

Hirai T, Jones EG (1989) A new parcellation of the human thalamus on the basis of histochemical staining. Brain Res Rev 14:1–34

Hirata H, Zakeshita S, Hu JW, Bereiter DA (2000) Cornea-responsive medullary dorsal horn neurons: modulation of local opioids and projections to thalamus and brain stem. J Neurophysiol 84:1050–1061

Hirschberg RM, Al-Chaer ED, Lawand NB, Westlund KN, Willis WD (1996) Is there a pathway in the posterior funiculus that signals visceral pain? Pain 67:291–305

Hogan EL, Krigman MR (1973) Herpes zoster myelitis: evidence for viral invasion of spinal cord. Arch Neurol 29:309–313

Hökfelt T (1991) Neuropeptides in perspective: the last ten years. Neuron 7:867–879

Hökfelt T, Zhang X, Wiesenfeld-Hallin Z (1994) Messenger plasticity in primary sensory neurons following axotomy and its functional implications. Trends Neurosci 17:22–30

Holland NR, Crawford TO, Hauer P, Cornblath DR, Griffin JW, McArthur, JC (1998) Small-fiber sensory neuropathies: clinical course and neuropathology of idiopathic cases. Ann Neurol 44:47–59

Hollman M, Heinemann S (1994) Cloned glutamate receptors. Annu Rev Neurosci 17:31–108

Hollmann M, O'Shea-Greenfield A, Rogers SW, Heinemann S (1989) Cloning by functional expression of a member of the glutamate receptor family. Nature 342:643–648

Holstege G, Kuypers HGJM (1982) The anatomy of brain stem pathways to the spinal cord in cat. A labeled amino acid tracing study. In: Kuypers HGJM, Martin GF (eds) Descending pathways to the spinal cord. Elsevier, Amsterdam, pp 145–175

Holzer P (1991) Capsaicin: cellular targets, mechanisms of action and selectivity for thin sensory neurons. Pharmacol Rev 43:143–201

Holzer P (1992) Peptidergic sensory neurons in the control of vascular functions: mechanisms and significance in the cutaneous and splanchnic vascular beds. Rev Physiol Biochem Pharmacol 121:49–146

Holzer P (1998) Neurogenic vasodilatation and plasma leakage in the skin. Gen Pharmacol 30:5–11

Holzer P, Maggi CA (1998) Dissociation of dorsal root ganglion neurons into afferent and efferent-like neurons. Neuroscience 86:389–398

Honore P, Luger NM, Sabino MA, Schwei MJ, Rogers SD, O'keefe PF, Ramnaraine ML, Clohisy DR, Mantyh PW (2000a) Osteoprotegerin blocks bone cancer-induced skeletal destruction, skeletal pain and pain-related neurochemical reorganization of the spinal cord. Nat Med 6:521–528

Honore P, Rogers SD, Schwei MJ, Salak-Johnson JL, Luger NM, Sabino MC, Clohisy DR, Mantyh PW (2000b) Murine models of inflammatory, neuropathic and cancer pain each generates a unique set of neurochemical changes in the spinal cord and sensory neurons. Neuroscience 98:585–598

Hope-Simpson RE (1965) The nature of herpes zoster: a long-term study and a new hypothesis. Proc Roy Soc Med 58:9–20

Hsieh JC, Belfrage M, Stone-Elander S, Hansson P, Ingvar M (1995) Central representation of chronic ongoing neuropathic pain studied by positron emission topography. Pain 63:225–236

Huber J, Grottel K, Mrowczynski W, Krutki P (1999) Spinoreticular neurons in the second sacral segment of the feline spinal cord. Neurosci Res 34:59–65

Hukkanen M, Konttinen YT, Rees RG, Santavirta S, Terenghi G, Polak JM (1992) Distribution of nerve endings and sensory neuropeptides in rat synovium, meniscus and bone. Int J Tissue React 14:1–10

Hudson AJ (2000) Pain perception and response: central nervous system mechanisms. Can J Neurol Sci 27:2–16

Hunt JR (1907) On herpetic inflammations of the geniculate ganglion: a new syndrome and its complications. J Nerv Ment Dis 34:73–96

Hunt JR (1937) Geniculate neuralgia (neuralgia of the nervus facialis). Arch Neurol Psychiat 37:253–285

Hunt SP, Mantyh PW (2001) The molecular dynamics of pain control. Nature Rev Neurosci 2:83–91

Hunt SP, Mantyh PW, Priestley JW (1992) The organization of biochemically characterized sensory neurons. In: Scott SA (ed) Sensory neurons: diversity, development and plasticity. Oxford University Press, New York, pp 60–76

Hwang SJ, Pagliardini S, Rustioni A, Valtschanoff JG (2001) Presynaptic kainate receptors in primary afferents to the superficial laminae of the rat spinal cord. J Comp Neurol 436:275–289

Hwang SJ, Burette A, Valtschanoff JG (2003) VR1-positive primary afferents contact NK1-positive spinoparabrachial neurons. J Comp Neurol 460:255–265

Hwang SJ, Burette A, Rustioni A, Valtschanoff JG (2004) Vanilloid receptor VR1-positive primary afferents are glutamatergic and contact spinal neurons that co-express neurokinin receptor NK1 and glutamate receptors. J Neurocytol 33:321–329

Hylden JLK, Hayashi H, Bennett GJ, Dubner R (1985) Spinal lamina I neurons projecting to the parabrachial area of the cat midbrain. Brain Res 336:195–198

Hyman RW, Ecker JR, Tenser RB (1983) Varicella-zoster virus RNA in human trigeminal ganglia. Lancet 2:814–816

Iadarola MJ, Berman KF, Zeffiro TA, Byas-Smith MG, Gracely RH, Max MB, Bennett GJ (1998) Neuronal activation during acute capsaicin-evoked pain and allodynia assessed with PET. Brain 121:931–947

Iggo A, Muir AR (1969) The structure and function of a slowly adapting touch corpuscle in hairy skin. J Physiol 200:763–796

Inoue K, Tsuda M, Koizumi S (2004) Chronic pain and microglia: the role of ATP. Novartis Found Symp 261:55–64

Isaac JTR, Nicoll RA, Malenka RC (1995) Evidence for silent synapses: implications for the expression of LTP. Neuron 15:427–434

Iwata K, Kenshalo DR, Dubner R, Nahin RL (1992) Diencephalic projections from the superficial and deep laminae of the medullary dorsal horn in the rat. J Comp Neurol 321:404–420

Jacquin MF, Barcia M, Rhoades RW (1989) Structure-function relationships in rat brainstem nucleus interpolaris. IV. Projection neurons. J Comp Neurol 282:45–62

Jacquin MF, Chiaia NL, Rhoades RW (1990) Trigeminal projections to contralateral dorsal horn: central extent, peripheral origins, and plasticity. Somatosens Mot Res 7:153–183

Jänig W (1996) The puzzle of "reflex sympathetic dystrophy": mechanisms, hypotheses, open questions. In: Jänig W, Stanton-Hicks M (eds) Reflex sympathetic dystrophy: a reappraisal. Progress in pain research and management, vol 6. IASP Press, Seattle, pp 1–24

Jessel T, Tsunoo A, Kanazawa I, Otsuka M (1979) Substance P: depletion in the dorsal horn of rat spinal cord after section of the peripheral processes of primary sensory neurons. Brain Res 168:247–259

Jessell TM, Yoshioka K, Jahr CE (1986) Amino acid receptor-mediated transmission at primary afferent synapses in rat spinal cord. J Exp Biol 124:239–258

Ji RR, Strichartz G (2004) Cell signalling and the genesis of neuropathic pain. Sci STKE 252: reE14

Ji RR, Woolf CJ (2001) Neuronal plasticity and signal transduction in nociceptive neurons: implications for the initiation and maintenance of pathological pain. Neurobiol Dis 8:1–10

Jones BE, Holmes CJ, Rodriguez-Veiga E, Mainville L (1991) GABA-synthesizing neurons in the medulla: their relationships to serotonin-containing and spinally projecting neurons in the rat. J Comp Neurol 313:349–367

Jones EG (1985) The thalamus. Plenum Press, New York

Jones EG (1997a) Thalamic organization and chemical anatomy. In: Steriade M, Jones EG, McCormick DA (eds) Thalamus, vol I. Organization and function. Elsevier, Amsterdam, pp 31–174

Jones EG (1997) A description of the human thalamus. In: Steriade M, Jones EG, McCormick DA (eds) Thalamus, vol II. Experimental and clinical aspects. Elsevier, Amsterdam, pp 425–499

Jones EG (1998) The thalamus of primates. In: Bloom FE, Björklund A, Hökfelt T (eds) Handbook of chemical neuroanatomy, vol 14. The primate nervous system, part II. Elsevier, Amsterdam, pp 1–298

Jones EG (2002a) Thalamic circuitry and thalamocortical synchrony. Philos Trans R Soc Lond B Biol Sci 357:1659–1673

Jones EG (2002b) Thalamic organization and function after Cajal. Prog Brain Res 136:333–357

Jones SL, Light AR (1990) Termination patterns of serotoninergic medullary raphespinal fibers in the rat lumbar spinal cord: an anterograde immunohistochemical study. J Comp Neurol 297:267–282

Jones SL, Light AR (1992) Serotoninergic medullary raphespinal projection to the lumbar spinal cord in the rat: a retrograde immunohistochemical study. J Comp Neurol 322:599–610

Ju G, Melander T, Ceccatelli S, Hökfelt T, Frey P (1987) Immunohistochemical evidence for a spinothalamic pathway co-containing cholecystokinin- and galanin-like immunoreactivities in the rat. Neuroscience 20:439–456

Julius D, Basbaum A (2001) Molecular mechanisms of nociception. Nature 413:203–210

Jung BF, Johnson RW, Griffin DR, Dworkin RH (2004) Risk factors for postherpetic neuralgia in patients with herpes zoster. Neurology 62:1545–1551

Kajander KC, Xu J (1995) Quantitative evaluation of calcitonin gene-related peptide and substance P levels in rat spinal cord following peripheral nerve injury. Neurosci Lett 186:184–188

Kakigi R, Tran TD, Qiu Y, Wang X, Nguyen TB, Inui K, Watanabe S, Hoshiyama M (2003) Cerebral responses following stimulation of unmyelinated C-fibers in humans: electro- and magneto-encephalographic study. Neurosci Res 45:255–275

Kalichman MW, Powell HC, Mizisin AP (1998) Reactive, degenerative, and proliferative Schwann cell responses in experimental galactose and human diabetic neuropathy. Acta Neuropathol 95:47–56

Kanazi G, Johnson RW, Dworkin RH (2000) Treatment of postherpetic neuralgia. An update. Drugs 59:1113–1126

Kanda M, Nagamine T, Ikeda A, Ohara S, Kunieda T, Fujiwara N, Yazawa S, Sawamoto N, Matsumoto R, Taki W, Shibasaki H (2000) Primary somatosensory cortex is actively involved in pain processing in human. Brain Res 853:282–289

Kapadia SE, LaMotte CC (1987) Deafferentation induced alterations in the rat dorsal horn. I. Comparison of peripheral nerve injury versus rhizotomy effects on presynaptic, postsynaptic and glial processes J Comp Neurol 266:183–197

Kapur D (2003) Neuropathic pain and diabetes. Diabetes Metab Res Rev 19 (Suppl 1):S9–S15

Katter JT, Burstein R, Giesler GJ (1991) The cells of origin of the spinohypothalamic tract in cats. J Comp Neurol 303:101–112

Katter JT, Dado RJ, Kostarczyk E, Giesler GJ (1996) Spinothalamic and spinohypothalamic tract neurons in the sacral spinal corn of rats. I. Location of antidromically identified axons in the cervical cord and diencephalon. J Neurophysiol 75:2581–2605

Kayalioglu G, Robertson B, Kristensson K, Grant G (1999) Nitric oxide synthase and interferon-gamma receptor immunoreactivities in relation to ascending spinal pathways to thalamus, hypothalamus, and the periaqueductal grey in the rat. Somatosens Mot Res 16:280–290

Keast JR, Stephensen TM (2000) Glutamate and aspartate immunoreactivity in dorsal root ganglion cells supplying visceral and somatic targets and evidence for peripheral axonal transport. J Comp Neurol 424:577–587

Kellenberger E, Hayat M (1991) Some basic concepts for the choice of methods. In: Hayat MA (ed) Colloidal gold: principles, methods and applications. Academic Press, San Diego, pp 1–30

Kellenberger E, Dürrenberger M, Villiger W, Carlemalm E, Wurtz M (1987) The efficiency of immunolabel on Lowicryl sections compared to theoretical predictions. J Histochem Cytochem 35:959–969

Kemplay SK, Webster KE (1986) A qualitative and quantitative analysis of the distributions of cells in the spinal cord and spinomedullary junction projecting to the thalamus of the rat. Neuroscience 17:769–789

Kemplay S, Webster KE (1989) A quantitative study of the projections of the gracile, cuneate and trigeminal nuclei and of the medullary reticular formation to the thalamus in the rat. Neuroscience 32:153–167

Kennedy PG, Grinfeld E, Gow J (1998) Latent varicella-zoster virus is located predominantly in neurons in human trigeminal ganglia. Proc Natl Acad Sci U S A 95:4658–4662

Kenshalo DR, Chudler EH, Anton F, Dubner R (1988) SI cortical nociceptive neurons participate in the encoding process by which monkeys perceive the intensity of noxious thermal stimulation. Brain Res 454:378–382

Kerr FWL (1975a) Neuroanatomical substrates of nociception in the spinal cord. Pain 1:325–336

Kerr FWL (1975b) The ventral spinothalamic tract and other ascending systems of the ventral funiculus of the spinal cord. J Comp Neurol 159:335–356

Kevetter GA, Willis WD (1982) Spinothalamic cells in the rat lumbar cord with collaterals to the medullary reticular formation. Brain Res 238:181–185

Kevetter GA, Willis WD (1983) Collaterals of spinothalamic cells in the rat. J Comp Neurol 215:453–464

Kevetter GA, Willis WD (1984) Collateralization in the spinothalamic tract: new methodology to support or deny phylogenetic theories. Brain Res Rev 7:1–14

Kevetter GA, Haber LH, Yezierski RP, Vhung JM, Martin RF, Willis WD (1982) Cells of origin of the spinoreticular tract in the monkey. J Comp Neurol 207:61–74

Khan GM, Chen SR, Pan HL (2002) Role of primary afferent nerves in allodynia caused by diabetic neuropathy in rats. Neuroscience 114:291–299

Kharazia VN, Weinberg RJ (1994) Glutamate and thalamic fibers terminating in layer IV of primary sensory cortex. J Neurosci 14:6021-6032

Kharazia VN, Wenthold RJ, Weinberg RJ (1996) GluR1-immunopositive interneurons in rat neocortex. J Comp Neurol 368:399–412

Kharazia VN, Phend KD, Rustioni A, Weinberg RJ (1996) EM colocalization of AMPA and NMDA receptor subunits at synapses in rat cerebral cortex. Neurosci Lett 210:37–40

Khasabov SG, Ghilardi JR, Mantyh PW, Simone DA (2005) Spinal neurons that express NK-1 receptors modulate descending controls that project through the dorsolateral funiculus. J Neurophysiol 93:998–1006

Kim SH, Chung JM (1992) An experimental model for peripheral neuropathy produced by segmental spinal nerve ligation in the rat. Pain 50:355–363

Kingery WS (1997) A critical review of controlled clinical trials for peripheral neuropathic pain and complex regional pain syndromes. Pain 73:123–139

Kitamura T, Yamada J, Sato H, Yamashita K (1993) Cells of origin of the spinoparabrachial fibers in the rat. J Comp Neurol 328:449–461

Kitamura T, Nagao S, Kunimoto K, Shirama K, Yamada J (2001) Cytoarchitectonic subdivisions of the parabrachial nucleus in the Japanese monkey (Macacus fuscatus) with special reference to spinoparabrachial fiber terminals. Neurosci Res 39:95–108

Klassen KP, Morton DR, Curtis GM (1951) the clinical physiology of the human bronchi. III. The effect of the vagus section on the cough reflex, bronchial calibre and clearance of bronchial secretions. Surgery 29:483–490

Kleinschmidt-DeMasters BK, Gilden DH (2001) Varicella-zoster virus infections of the nervous system. Clinical and pathologic correlates. Arch Pathol Lab Med 125:770–780

Kleinschmidt-DeMasters BK, Amlie-Lefond C, Gilden DH (1996) The pattern of varicella zoster virus encephalitis. Hum Pathol 27:927–938

Klop EM, Mouton LJ, Holstege G (2004a) How many spinothalamic tract cells are there? A retrograde tracing study in cat. Neurosci Lett 360:121–124

Klop EM, Mouton LJ, Holstege G (2004b) Less than 15% of the spinothalamic fibers originate from neurons in lamina I in cat. Neurosci Lett 360:125–128

Koerber HR, Mirnics K, Kavookjian AM, Light AR (1999) Ultrastructural analysis of ectopic synaptic boutons arising from peripherally regenerated primary afferent fibers. J Neurophysiol 81:1636–1644

Kohama I, Ishikawa K, Kocsis JD (2000) Synaptic reorganization in the substantia gelatinosa after peripheral nerve neuroma formation: aberrant innervation of lamina II neurons by Abeta afferents. J Neurosci 20:1538–1549

Koltzenburg M, Torebjörk HE, Wahren LK (1994) Nociceptor modulated central sensitization causes mechanical hyperalgesia in acute chemogenetic and chronic neuropathic pain. Brain 117:579–591

Kondo E, Kiyama H, Yamano M, Shida T, Ueda Y, Tohyama M (1995) Expression of glutamate (AMPA type) and γ-aminobutyric acid (GABA)A receptors in the rat caudal trigeminal spinal nucleus. Neurosci Lett 186:169–172

Kost RG, Straus SE (1996) Postherpetic neuralgia: pathogenesis, treatment, and prevention. N Engl J Med 335:32–42

Kostarczyk E, Zhang X, Giesler GJ (1997) Spinohypothalamic tract neurons in the cervical enlargement of rats: location of antidromically identified ascending axons and their collateral branches in the contralateral brain. J Neurophysiol 77:435–451

Kress M, Fickenscher H (2001) Infection by human varicella-zoster virus confers norepinephrine sensitivity to sensory neurons from rat dorsal root ganglia. FASEB J 15:1037–1043

Kruger L (1988) Morphological features of thin sensory afferent fibers: a new interpretation of "nociceptor" function. Prog Brain Res 74:253–257

Kruger L (1996) The functional morphology of thin sensory axons: some principles and problems. Prog Brain Res 113:255–272

Kruger L, Sampogna SL, Rodin BE, Clague J, Brecha N, Yeh Y (1985) Thin-fiber cutaneous innervation and its intraepidermal contribution studied by labelling methods and neurotoxin treatment in rats. Somatosens Res 2:335–356

Kruger L, Perl ER, Sedivec MJ (1981) Fine structure of myelinated nociceptor endings in cat hairy skin. J Comp Neurol 198:137–154

Kruger L, Kavookjan AM, Kumazawa T, Light AR, Mizumura K (2003a) Nociceptor structural specialization in canine and rodent testicular "free" nerve endings. J Comp Neurol 463:197–211

Kruger L, Light AR, Schweizer FE (2003b) Axonal terminals of sensory neurons and their morphological diversity. J Neurocytol 32:205–216

Kus L, Sanderson JJ, Beitz AJ (1995) N-Methyl-D-aspartate R1 messenger RNA and [^{125}I]MK-801 binding decreases in rat spinal cord after unilateral hind paw inflammation. Neuroscience 68:159–165

Kuypers HGJM, Bentivoglio M, Catsman-Berrevoets CE, Bharos TB (1980) Double retrograde neuronal labelling through divergent axon collaterals using two fluorescent tracers with the same excitation wavelength which label different features of the cell. Exp Brain Res 40:383–392

Kwiat G, Basbaum A (1990) Organization of tyrosine hydroxylase- and serotonin-immunoreactive brainstem neurons with axon collaterals to the periaqueductal gray and the spinal cord. Brain Res 528:83–94

Kyrozis A, Goldstein PA, Heath MJS, MacDermott AB (1995) Calcium entry through a subpopulation of AMPA receptors desensitized neighbouring NMDA receptors in rat dorsal horn neurons. J Physiol (Lond) 485:373–381

LaGuardia JJ, Cohrs RJ, Gilden DH (2000) Numbers of neurons and non-neuronal cells in human trigeminal ganglia. Neurol Res 22:565–566

Lahuerta J, Bowsher D, Lipton S, Buxton PH (1994) Percutaneous cervical cordotomy: a review of 181 operations on 146 patients with a study on the location of "pain fibers" in the C-2 spinal cord segment of 29 cases. J Neurosurg 80:975–985

LaMotte CC, Kapadia SE, Shapiro CM (1991) Central projections of the sciatic, saphenous, median, and ulnar nerves of the rat demonstrated by transganglionic transport of choleragenoid-HRP (B-HRP) and wheat germ agglutinin-HRP (WGA-HRP). J Comp Neurol 311:546–562

LaMotte CC, Kapadia SE (1993) Deafferentation-induced terminal field expansion of myelinated saphenous afferents in the adult rat dorsal horn and the nucleus gracilis following pronase injection of the sciatic nerve. J Comp Neurol 330:83–94

LaMotte CC, Arsenault KE, Wolfe MA, Helgren ME, Kapadia SE (1996) Deafferentation-induced alterations of receptor density in the dorsal horn. Soc Neurosci Abstr 22:860

Langford LA, Coggeshall RE (1984) Branching of sensory axons in the peripheral nerve of the rat. J Comp Neurol 203:745–750

La Vail JH, Johnson WE, Spencer LC (1993) Immunohistochemical identification of trigeminal ganglion neurons that innervate the mouse cornea: relevance to intercellular spread of herpes simplex virus. J Comp Neurol 327:133–140

Laurberg S, Sorensen KE (1985) Cervical dorsal root ganglion cells with collaterals to both shoulder skin and the diaphragm. A fluorescent double labelling study in the rat. A model for referred pain? Brain Res 331:160–163

Lawand NB, Willis WD, Westlund KN (1997) Excitatory amino acid receptor involvement in peripheral nociceptive transmission in rats. Eur J Pharmacol 324:169–174

Lawson SN (1992) Morphological and biochemical cell types of sensory neurons. In: Scott SA (ed) Sensory neurons: diversity, development and plasticity. Oxford University Press, New York, pp 27–59

Lawson SN (2002) Phenotype and function of somatic primary afferent nociceptive neurones with C-, Adelta- or Aalpha/beta-fibres. Exp Physiol 87:239–244

Lawson SN, Waddell PJ, McCarthy PW (1987) A comparison of the electrophysiological and immunocytochemical properties of rat dorsal root ganglion neurons with A and C-fibers. In: Schmidt RF, Schaible HG, Vahle-Hinz C (eds) Fine afferent nerve fibers and pain. VCH Publishers, Weinheim and New York, pp 193–203

Lawson SN, Crepps BA, Perl ER (1997) Relationship of substance P to afferent characteristics of dorsal root ganglion neurones in guinea-pig. J Physiol 505:177–191

Lawson SN, Crepps BA, Perl ER (2002) Calcitonin gene-related peptide immunoreactivity and afferent receptive properties of dorsal root ganglion neurones in guinea-pigs. J Physiol 540:989–1002

Lazarov NE (2002) Comparative analysis of the chemical neuroanatomy of the mammalian trigeminal ganglion and mesencephalic trigeminal nucleus. Prog Neurobiol 66:19–60

Lazarov N, Dandov A (1998) Distribution of NADPH-diaphorase and nitric oxide synthase in the trigeminal ganglion and mesencephalic trigeminal nucleus of the cat. A histochemical and immunohistochemical study. Acta Anat 163:191–200

LeBars D, Dickenson AH, Besson JM (1979a) Diffuse noxious inhibitory controls (DNIC). I. Effects on dorsal horn convergent neurons in the rat. Pain 6:283–304

LeBars D, Dickenson AH, Besson JM (1979b) Diffuse noxious inhibitory controls (DNIC). II. Lack of effect on non-convergent neurons, supraspinal involvement and theoretical implications. Pain 6:305–327

Lee JH, Price RH, Williams FG, Mayer B, Beitz AJ (1993) Nitric oxide synthase is found in some spinothalamic neurons and in neuronal processes that appose spinal neurons that express Fos induced by noxious stimulation. Brain Res 608:324–333

Leijon G, Boivie J, Johannson I (1989) Central post-stroke pain: neurologic symptoms and pain characteristics. Pain 36:13–25

Lenz FA, Gracely RH, Zirh A, Romanoski AJ, Dougherty PM (1997) The sensory-limbic model of pain memory. Pain Forum 6:22–31

Lenz FA, Lee J-I, Garoznik I-M, Rowland LH, Dougherty PM, Hua SE (2000) Human thalamus reorganization related to nervous system injury and dystonia. Prog Brain Res 129:259–273

Levine JD, Fields HL, Basbaum AL (1993) Peptides and the primary afferent nociceptor. J Neurosci 13:2273–2286

Lewin GR, Moshourab R (2004) Mechanosensation and pain. J Neurobiol 61:30–44

Lewis GW (1958) Zoster sine herpete. Br Med J 2:418–421

Li JL, Kaneko T, Shigemoto R, Mizuno N (1997) Distribution of trigeminohypothalamic and spinohypothalamic tract neurons displaying substance P receptor-like immunoreactivity in the rat. J Comp Neurol 378:508–521

Li JL, Kaneko T, Mizuno N (1999) Preprodynorphin-like immunoreactivity in medullary dorsal horn neurons projecting to the thalamic regions in the rat. Neurosci Lett 264:13–16

Li W, Neugebauer V (2004) Differential roles of mGluR1 and mGluR5 in brief and prolonged nociceptive processing in central amygdala neurons. J Neurophysiol 91:13–24

Li YQ (1999) Substance P receptor-like immunoreactive neurons in the caudal spinal trigeminal nucleus send axons to the gelatinosus thalamic nucleus in the rat. J Hirnforsch 39:277–282

Li YQ, Li H, Kaneko T, Mizuno N (1999) Substantia gelatinosa neurons in the medullary dorsal horn: an intracellular labeling study in the rat. J Comp Neurol 411:399–412

Li YQ, Li H, Yang K, Kaneko T, Mizuno N (2000a) Morphologic features and electrical membrane properties of projection neurons in the marginal layer of the medullary dorsal horn of the rat. J Comp Neurol 424:24–36

Li YQ, Li H, Yang K, Wang ZM, Kaneko T, Mizuno N (2000b) Intracellular labelling study of neurons in the superficial part of the magnocellular layer of the medullary dorsal horn of the rat. J Comp Neurol 428:641–655

Light AR (1992) The initial processing of pain and its descending control: spinal and trigeminal systems. Karger, Basel

Light AR, Perl ER (1979a) Reexamination of the dorsal root projection to the spinal dorsal horn including observations on the differential termination of coarse and fine fibers. J Comp Neurol 186:117–132

Light AR, Perl ER (1979b) Spinal terminations of functionally identified primary afferent neurons with slowly conducting myelinated fibers. J Comp Neurol 186:133–150

Light AR, Trevino DL, Perl ER (1979) Morphological features of functionally defined neurons in the marginal zone and substantia gelatinosa of the spinal dorsal horn. J Comp Neurol 186:151–172

Light A, Sedivec M, Casale E, Jones S (1993) Physiological and morphological characteristics of spinal neurons projecting to the parabrachial region of the cat. Somatosens Mot Res 10:309–325

Lilie HM, Wassilew S (2003) The role of antivirals in the management of neuropathic pain in the older patient with herpes zoster. Drugs Aging 20:561–570

Lima D, Almeida A (2002) The medullary dorsal reticular nucleus as a pronociceptive centre of the pain control system. Prog Neurobiol 66:81–108

Lima D, Coimbra A (1988) The spinothalamic system of the rat: structural types of retrogradely labelled neurons in the marginal zone (lamina I). Neuroscience 27:215–230

Lima D, Coimbra A (1991) Neurons in the substantia gelatinosa Rolandi (lamina II) project to the caudal ventrolateral reticular formation of the medulla oblongata in the rat. Neurosci Lett 132:16–18

Lima D, Mendes-Ribeiro JA, Coimbra A (1991) The spino-latero-reticular system of the rat: projections from the superficial dorsal horn and structural characterization of the neurons involved. Neuroscience 45:137–152

Liu CN, Chambers WW (1958) Intraspinal sprouting of dorsal root axons. Arch Neurol 79:46–61

Liu H, Wang H, Sheng Jan LY, Jan YN, Basbaum AI (1994) Evidence for presynaptic N-methyl-D-aspartate autoreceptors in the spinal cord dorsal horn. Proc Natl Acad Sci U S A 91:8383–8387

Liu XG, Sandkühler J (1995) Long-term potentiation of C-fiber-evoked potentials in the rat spinal dorsal horn is prevented by spinal N-methyl-D-aspartic acid receptor blockage. Neurosci Lett 191:43–46

Livengood JM (2000) The role of stress in the development of herpes zoster and postherpetic neuralgia. Curr Rev Pain 4:219–226

Lojeski E, Stevens RA (2000) Postherpetic neuralgia in the cancer patient. Curr Rev Pain 4:219–226

Lu CR, Hwang SJ, Phend KD, Rustioni A, Valtschanoff JG (2002) Primary afferent terminals in spinal cord express presynaptic AMPA receptors. J Neurosci 22:9522–9529

Lund RD, Webster KE (1967a) Thalamic afferents from the dorsal column nuclei. An experimental anatomical study in the rat. J Comp Neurol 130:301–312

Lund RD, Webster KE (1967b) Thalamic afferents from the spinal cord and trigeminal nuclei. J Comp Neurol 130:313–328

Lungu O, Annunziato P, Gerschon A, Staugaitis SM, Josefson D, LaRussa P, Silverstein SJ (1995) Reactivated and latent varicella-zoster virus in human dorsal root ganglia. Proc Natl Acad Sci U S A 85:9773–9777

Lungu O, Panagiotidis CA, Annunziato PW, Gershon AA, Silverstein SJ (1998) Aberrant intracellular localization of Varicella-Zoster virus regulatory proteins during latency. Proc Natl Acad Sci U S A 95:7080–7085

Luo H, Cui S, Chen D, LiuJ, Liu Z (2004) Immunohistochemical detection of islet-1 and neuronal nitric oxide synthase in the dorsal root ganglia (DRG) of sheep fetuses during gestation. J Histochem Cytochem 52:797–803

Luo ZD, Chaplan SR, Scott BP, Cizkova D, Calcutt NA, Yaksh TL (1999) Neuronal nitric oxide synthase mRNA upregulation in rat sensory neurons after spinal nerve ligation: lack of a role in allodynia development. J Neurosci 19:9201–9208

Ma W, Bisby MA (1998) Partial and complete sciatic nerve injuries induce similar increases of neuropeptide Y and vasoactive intestinal polypeptide immunoreactivities in primary sensory neurons and their central projections. Neuroscience 86:1217–1234

Mach DB, Rogers SD, Sabino MAC, Luger NM, Schwei MJ, Pomonis JD, Keyser CP, Clohisy DR, Adams DJ, O'Leary P, Mantyh PW (2002) Origins of skeletal pain: sensory and sympathetic innervation of the mouse femur. Neuroscience 113:156–166

Madiai F, Hussain SR, Goettl VM, Burry RW, Stephens RL Jr, Hackshaw KV (2002) Upregulation of FGF-2 in reactive spinal cord astrocytes following unilateral lumbar spinal cord ligation. Exp Brain Res 148:366–376

Magnusson KR, Larson AA, Madl JE, Altschuler RA, Beitz AJ (1986) Co-localization of fixative-modified glutamate and glutaminase in neurons of the spinal trigeminal nucleus of the rat: an immunohistochemical and immunoradiochemical analysis. J Comp Neurol 247:477–490

Magnusson KR, Clements JR, Larson AA, Madl JE, Beitz AJ (1987) Localization of glutamate in trigeminothalamic projection neurons: a combined retrograde transport-immunohistochemical study. Somatosens Res 4:177–190

Mahalingam R, Wellish M, Wolf W, Dueland AN, Cohrs R, Vafai A, Gilden D (1990) Latent varicella-zoster viral DNA in human trigeminal and thoracic ganglia. N Engl J Med 323:627–631

Mahalingam R, Kennedy PG, Gilden DH (1999) The problems of latent varicella- zoster virus in human ganglia: precise cell location and viral content. J Neurovirol 5:445–448

Mai JK, Assheuer J, Paxinos G (1997) Atlas of the human brain. Academic Press, San Diego

Mailis A, Furlan A (2003) Sympathectomy for neuropathic pain. Cochrane Database Syst Rev 2: CD002918

Majewski M, Sienkiewicz W, Kaleczyc J, Mayer B, Czaja K, Lakomy M (1995) The distribution and co-localization of immunoreactivity to nitric oxide synthase, vasoactive intestinal polypeptide and substance P within nerve fibres supplying bovine and porcine female genital organs. Cell Tissue Res 281:445–464

Malick A, Strassman RM, Burstein R (2000) Trigeminohypothalamic and reticulohypothalamic tract neurons in the upper cervical spinal cord and caudal medulla of the rat. J Neurophysiol 84:2078–2112

Manfredi PL, Gonzales GR, Sady R, Chandler S, Payne R (2003) Neuropathic pain in patients with cancer. J Palliat Care 19:115–118

Mannion RJ, Woolf CJ (2000) Pain mechanisms and management: a central perspective. Clin J Pain 16 [Suppl 3]:S144–S156

Mannion RJ, Doubell TP, Coggeshall RE, Woolf CJ (1996) Collateral sprouting of uninjured primary afferent A-fibres into the superficial dorsal horn of the adult rat spinal cord after topical capsaicin treatment to the sciatic nerve. J Neurosci 16:5189–5195

Mannion RJ, Doubell TP, Gill H, Woolf CJ (1998) Deafferentation is insufficient to induce sprouting of A-fibre central terminals in the rat dorsal horn. J Comp Neurol 393:135–144

Mannion RJ, Costigan M, Decorsterd I, Amaya F, Ma QP, Holstege JC, Ji RR, Acheson A, Lindsay RM, Wilkinson GA, Woolf CJ (1996) Neurotrophins: peripherally and centrally acting modulators of tactile stimulus-induced inflammatory pain hypersensitivity. Proc Natl Acad Sci U S A 96:9385–9390

Manola L, Roelofsen BH, Holsheimer J, Marani E, Geelen J (2005) Modelling motor cortex stimulation for chronic pain control: electrical potential filed, activating functions and responses of simple nerve fiber models. Med Biol Eng Comput 43:335–343

Mantle-St John LA, Tracey DJ (1987) Somatosensory nuclei in the brainstem of the rat: independent projections to the thalamus and cerebellum. J Comp Neurol 255:259–271

Mantyh PW (1982) The ascending input to the midbrain periaqueductal gray of the primate. J Comp Neurol 211:50–64

Mantyh PW (1983) The spinothalamic tract in the primate: a re-examination using wheatgerm agglutinin conjugated to horseradish peroxidase. Neuroscience 9:847–862

Mantyh PW, Hunt SP (2004) Setting the tone: superficial dorsal horn projection neurons regulate pain sensitivity. Trends Neurosci 27:582–584

Mantyh PW, Clohisy DR, Koltzenburg M, Hunt SP (2002) Molecular mechanisms of cancer pain. Nature Rev Cancer 2:201–209

Mantyh PW, Hunt SP (2004) Mechanisms that generate and maintain bone cancer pain. Novartis Found Symp 260:221–238

Marani E, Schoen JHR (2005) A reappraisal of the ascending systems in Man with emphasis on the medial lemniscus. Adv Anat Embryol Cell Biol 182:1–87

Marchettini P, Simone DA, Caputi G, Ochoa JL (1996) Pain of excitation of identified muscle nociceptors in humans. Brain Res 740:109–116

Maren S, Tocco G, Standley S, Baudry M, Thompson RF (1993) Postsynaptic factors in the expression of long-term potentiation (LTP): increased glutamate receptor binding following LTP induction in vivo. Proc Natl Acad Sci U S A 90:9654–9658

Marfurt CF, Echtenkamp SF (1988) Central projections and trigeminal ganglion location of corneal afferent neurons in the monkey, Macaca fascicularis. J Comp Neurol 272:370–382

Marfurt CF, Rajchert DM (1991) Trigeminal primary afferent projections to "non-trigeminal" areas of the rat central nervous system. J Comp Neurol 303:489–511

Marfurt CF, Kingsley RE, Echtenkamp SF (1989) Sensory and sympathetic innervation of the mammalian cornea. A retrograde tracing study. Invest Ophtalmol Vis Sci 30:461–472

Marfurt CF, Murphy CJ, Florczak JL (2001) Morphology and neurochemistry of canine corneal innervation. Invest Ophtamol Vis Sci 42:2242–2251

Marshall GE, Shehab SA, Spike RC, Todd AJ (1996) Neurokinin-1 receptors on lumbar spinothalamic neurons in the rat. Neuroscience 72:255–263

Matsubara A, Laake JH, Davanger S, Usami S, Ottersen OP (1996) Organization of AMPA receptor subunits at a glutamate synapse: a quantitative immunogold analysis of hair cell synapses in the rat organ of Corti. J Neurosci 16:4457–4467

McCarson KE, Krause JE (1994) NK-1 and NK-3 type tachykinin receptor mRNA expression in the rat spinal cord dorsal horn is increasing during adjuvant or formalin-induced nociception. J Neurosci 14:712–720

McCleskey EW, Gold MS (1999) Ion channels of nociception. Annu Rev Physiol 61:835–856

McGonigle DJ, Maxwell DJ, Shehab SAS, Kerr R (1996) Evidence for the presence of neurokinin-1 receptors on dorsal horn spinocerebellar tract cells in the rat. Brain Res 742:1–9

McHugh JM, McHugh WB (2000) Pain: neuroanatomy, chemical mediators, and clinical implications. AACN Clin Issues 11:168–178

McHugh JM, McHugh WB (2004) Diabetes and peripheral sensory neurons: what we don't know and how it can hurt us. AACN Clin Issues 15:136–149

McLachlan EM, Jänig W, Devor M, Michaelis M (1993) Peripheral nerve injury triggers noradrenergic sprouting within dorsal root ganglia. Nature 363:543–546

McMahon SB, Bennett DLH (1999) Trophic factors and pain. In: Wall PD, Melzack R (eds) Textbook of pain. Churchill-Livingstone, New York, pp 105–128

McMahon SB, Kett-White R (1991) Sprouting of peripherally regenerating primary sensory neurones in the adult central nervous system. J Comp Neurol 304:307–315

McMahon SB, Koltzenburg M (1994) Silent afferents and visceral pain. In: Pharmacological approaches to the treatment of chronic pain: new concepts and critical issues. Progress in pain research and management, vol 1. IASP Press, Seattle, pp 11–30

McMahon SB, Lewin GR, Wall PD (1993) Central hyperexcitability triggered by noxious input. Curr Opin Neurobiol 3:602–610

Mehler WR (1962) The anatomy of the so-called "pain tract" in man: an analysis of the course and distribution of the ascending fibers of the fasciculus anterolateralis. In: French JD, Porter RW (eds) Basic research in paraplegia. Springfield, Illinois, pp 26–55

Mehler WR (1966) The posterior thalamic region in man. Confin Neurol 27:18–29

Mehler WR, Feferman ME, Nauta WJH (1960) Ascending axon degeneration following anterolateral cordotomy: an experimental study in the monkey. Brain 83:718–751

Meier JL, Holman RP, Croen KD, Smialek JE, Straus SE (1993) Varicella-zoster virus transcription in human trigeminal ganglia. Virology 193:193–200

Meldrum BS (2000) Glutamate as a neurotransmitter in the brain: review of physiology and pathology. J Nutr 130:1007S-1015S

Meller ST, Gebhart GF (1992) A critical review of the afferent pathways and the potential chemical mediators involved in cardiac pain. Neuroscience 48:501–524

Melzack R, Wall PD (1965) Pain mechanisms: a new theory. Science 150:971–979

Menetrey D, Chaouch A, Besson JM (1980) Location and properties of dorsal horn neurons at origin of spinoreticular tract in lumbar enlargement of the rat. J Neurophysiol 44:862–877

Menetrey D, Chaouch A, Binder D, Besson JM (1982) The origin of the spinomesencephalic tract in the rat: an anatomical study using the retrograde transport of horseradish peroxidase. J Comp Neurol 206:193–207

Menetrey D, Roudier F, Besson JM (1983) Spinal neurons reaching the lateral reticular nucleus as studied in the rat by retrograde transport of horseradish peroxidase. J Comp Neurol 220:439–452

Menetrey D, Gannon A, Levine JD, Basbaum AI (1989) Expression of c-fos protein in interneurons and projection neurons of the rat spinal cord in response to noxious somatic, articular, and visceral stimulation. J Comp Neurol 285:177–195

Mense S (1996) Nociceptors in skeletal muscle and their reaction to pathological tissue changes. In: Belmonte C, Cervero F (eds) Neurobiology of nociceptors. Oxford University Press, Oxford, pp 184–201

Merighi A, Polak JM, Theodosis TD (1991) Ultrastructural visualization of glutamate and aspartate immunoreactivities in the rat dorsal horn, with special reference to the co-localization of glutamate, substance P and calcitonin-gene related peptide. Neuroscience 40:67–80

Messlinger K (1996) Functional morphology of nociceptive and other fine sensory endings (free nerve endings) in different tissues. Prog Brain Res 113:273–298

Mesulam MM, Mufson EJ, Wainer BH, Levey AI (1984) Central cholinergic pathway in the rat: An overview based on alternative nomenclature (Ch1-Ch6). Neuroscience 10:1185–1201

Mesulam MM, Geula C, Bothwell MA, Hersh LB (1989) Human reticular formation: cholinergic neurons of the pedunculopontine and laterodorsal tegmental nuclei and some cytochemical comparisons to forebrain cholinergic neurons. J Comp Neurol 281:611–633

Micevych PE, Kruger L (1992) The status of calcitonin gene-related peptide as an effector peptide. Ann N Y Acad Sci 657:379–396

Michael GJ, Priestly JV (1999) Differential expression of the mRNA for the vanilloid receptor subtype 1 in cells of the adult rat dorsal root and nodose ganglia and its downregulation by axotomy. J Neurosci 19:1844–1854

Miki K, Fukuoka T, Tokunaga A, Noguchi K (1998) Calcitonin gene-related peptide increase in rat spinal cord and dorsal column nucleus following peripheral nerve injury: up-regulation in a subpopulation of primary afferent neurons. Neuroscience 82:1243–1252

Millan, MJ (1999) The induction of pain: an integrative review. Prog Neurobiol 57:1–164

Mille-Hamard L, Bauchet L, Baillet-Derbin C, Horvat JC (1999) Estimation of the number and size of female adult rat C4, C5 and C6 dorsal root ganglia (DRG) neurons. Somatosens Mot Res 16:223–238

Milligan ED, O'Connor KA, Nguyen KT, Armstrong CB, Twinning C, Gaykema RP, Holguin A, Martin D, Maier SF, Watkins LR (2001) Intrathecal HIV-1 envelope glycoprotein gp120 enhanced pain states mediated by spinal cord proinflammatory cytokines. J Neurosci 21:2808–2819

Milligan ED, Twinning C, Chacur M, Biedenkapp J, O'Connor K, Poole S, Tracey K, Martin D, Maier SF, Watkins LR (2003) Spinal glia and proinflammatory cytokines mediate mirror-image neuropathic pain in rats. J Neurosci 23:1026–1040

Milligan ED, Zapata V, Chacur M, Schoeniger D, Biedenkapp J, O'Connor KA, Verge GM, Chapman G, Green P, Foster AC, Naeve GS, Maier SF, Watkins LR (2004) Evidence that exogenous and endogenous fractalkine can induce spinal nociceptive facilitation in rats. Eur J Neurosci 20:2294–2302

Minghetti L, Levi G (1998) Microglia as effector cells in brain damage and repair: focus on prostanoids and nitric oxide. Prog Neurobiol 55:1–26

Mizisin AP, Shelton GD, Wagner S, Rusbridge C, Powell HC (1998) Myelin splitting, Schwann cell injury and demyelination in feline diabetic neuropathy. Acta Neuropathol 95:171–174

Moalem G, Xu K, Yu L (2004) T lymphocytes play a role in neuropathic pain following peripheral nerve injury in rats. Neuroscience 129:767–777

Molander C, Grant G (1995) Spinal cord cytoarchitecture: In Paxinos G (ed) The rat nervous system, 2nd edn. Academic Press, San Diego, pp 39–45

Moore KA, Baba H, Woolf CJ (2000) Synaptic transmission and plasticity in the superficial dorsal horn. Prog Brain Res 129:63–80

Montoya P, Ritter K, Huse E, Larbig W, Braun C, Topfner S, Lutzenberger W, Grodd W, Flor H, Birbaumer N (1998) The cortical somatotopic map and phantom phenomena in subjects with congenital limb atrophy and traumatic amputees with phantom limb pain. Eur J Neurosci 10:1095–1102

Morris R, Southam E, Braid DJ, Gartwaite J (1992) Nitric oxide may act as a messenger between dorsal root ganglion neurones and their satellite cells. Neurosci Lett 137:29–32

Morris R, Cheunsuang O, Stewart A, Maxwell D (2004) Spinal dorsal horn neurone targets for nociceptive primary afferents: do single neurone morphological characteristics suggest how nociceptive information is processed at the spinal level. Brain Res Rev 46:173–190

Müller LJ, Marfurt CF, Kruse F, Tervo TMT (2003) Corneal nerves: structure, contents and function. Exp Eye Res 76:521–542

Munger BL, Halata Z (1983) The sensory innervation of the primate facial skin. I. Hairy skin. Prog Brain Res 5:45–80

Munger BL, Ide C (1988) The structure and function of cutaneous sensory receptors. Arch Histol Cytol 51:1–34

Nagashima K, Nakazawa M, Endo H (1975) Pathology of the human spinal ganglia in varicella-zoster virus infection. Acta Neuropathol 33:105–117

Nagy I, Dray A, Urban L (1995) Possible branching of myelinated primary afferent fibres in the dorsal root ganglion of the rat. Brain Res 703:223–226

Nahin RL (1987) Immunocytochemical identification of long ascending peptidergic neurons contributing to the spinoreticular tract in the rat. Neuroscience 23:859–869

Nair PN (1995) Neural elements in dental pulp and dentin. Oral Surg Oral Med Oral Pathol Oral Radiol Endod 80:710–719

Nakamura S, Myers RR (1999) Myelinated afferents sprout into lamina II of L3–5 dorsal horn following chronic constriction nerve injury in rats. Brain Res 818:285–290

Narita M, Yajima Y, Aoki T, Ozaki S, Mizuguchi H, Tseng LF, Suzuki T (2000) Up-regulation of the TrkB receptor in mice injured by the partial ligation of the sciatic nerve. Eur J Pharmacol 401:187–190

Narita M, Oe K, Kato H, Shibasaki M, Narita M, Yajima Y, Yamazaki M, Suzuki T (2004) Implication of spinal protein kinase C in the suppression of morphine-induced rewarding effect under a neuropathic pain-like state in mice. Neuroscience 125:545–551

Nathan PW, Smith M (1979) Clinico-anatomical correlation in anterolateral chordotomy. In: Bonica JJ (ed) Advances in pain research and therapy. Raven Press, New York, pp 921–926

Nathan PW, Smith M, Deacon P (2001) The crossing of the spinothalamic tract. Brain 124:793–803

Nauta HJW, Soukup VM, Fabian RH, Lin JT, Grady JJ, Williams CG, Campbell GA, Westlund KN, Willis WD (2000) Punctate midline myelotomy for the relief of visceral cancer pain. J Neurosurg 92 [Suppl 2]:125–130

Newman HM, Stevens RT, Apkarian AV (1996) Direct spinal projections to limbic and striatal areas: anterograde transport studies from the upper cervical spinal cord and the cervical enlargement in squirrel monkey and rat. J Comp Neurol 365:640–658

Nicholajsen L, Jensen TS (2001) Phantom limb pain. Br J Anaesth 87:107–116

Noguchi K, Senba E, Morita Y, Sato M, Tohyama M (1989) PreproVIP and preprotachykinin mRNAs in the rat dorsal root ganglion cells following peripheral axotomy. Mol Brain Res 6:327–330

Noguchi K, Senba E, Morita Y, Sato M, Tohyama M (1990) Alpha-CGRP and beta-CGRPmRNAs are differentially regulated in the rat spinal cord and dorsal root ganglion. Mol Brain Res 7:299–304

Nurmikko T, Wells C, Bowsher D (1991) Pain and allodynia in postherpetic neuralgia: role of somatic and sympathetic nervous systems. Acta Neurol Scand 84:146–152

Nusser Z, Mulvihill E, Streit P, Somogyi P (1994) Subsynaptic segregation of metabotropic and ionotropic glutamate receptors as revealed by immunogold localization. Neuroscience 61:421–427

Nusser Z, Roberts JD, Baude A, Richards JG, Somogyi P (1995a) Relative densities of synaptic and extrasynaptic GABAA receptors on cerebellar granule cells as determined by a quantitative immunogold method. J Neurosci 15:2948–2960

Nusser Z, Roberts JDB, Baude A, Richards JG, Sieghart W, Somogyi P (1995b) Immunocytochemical localization of the a1 and b2/3 subunits of the GABAA receptor in relation to specific GABAergic synapses in the dentate gyrus. Eur J Neurosci 7:630–646

Oaklander AL (1999) The pathology of shingles: Head and Campbell's 1900 monograph. Arch Neurol 56:1292–1294

Oaklander AL, Romans K, Horasek S, Stocks A, Hauer P, Meyer RA (1998) Unilateral postherpetic neuralgia is associated with bilateral neuron damage. Ann Neurol 44:789–795

O'Brien C, Woolf CJ, Fitzgerald M, Lindsay RM, Molander C (1989) Differences in the chemical expression of rat primary afferent neurons which innervate skin, muscle or joint. Neuroscience 32:493–502

Ochoa JL (1999) Truths, errors, and lies around "reflex sympathetic dystrophy" and "complex regional pain syndrome". J Neurol 246:875–879

Ochoa J, Verdugo RJ (2001) Mechanisms of neuropathic pain: nerve, brain and psyche: perhaps the dorsal horn but not the sympathetic system. Clin Auton Res 11:335–339

O'Connell JX, Nanthakumar SS, Nielsen GP, Rosenberg AE (1998) Osteoid osteoma: the uniquely innervated bone tumour. Modern Pathol 11:175–180

Olszewski J, Baxter D (1954) Cytoarchitecture of the human brain stem. Karger, Basel

Oshiro Y, Fuijita N, Tanaka H, Hirabuki N, Nakamura H, Yoshiya I (1998) Functional mapping of pain-related activation with echoplanar MRI: significance of SII-insular region. Neuroreport 9:2285–2289

Ottersen OP (1989) Quantitative electron microscopic immunocyto-chemistry of neuroactive amino acids. Anat Embryol 180:1–15

Pagni C (1998) Central pain. A neurosurgical challenge. Edizioni Minerva Medica, Torino

Palecek J, Willis WD (2003) The dorsal column pathway facilitates visceromotor responses to colorectal distension after colon inflammation in rats. Pain 104:501–507

Palecek J, Paleckova V, Willis WD (2002) The roles of pathways in the spinal cord lateral and dorsal funiculi in signaling nociceptive somatic and visceral stimuli in rats. Pain 96:297–307

Palecek J, Paleckova V, Willis WD (2003a) Postsynaptic dorsal column neurons express NK1 receptors following colon inflammation. Neuroscience 116:565–572

Palecek J, Paleckova V, Willis WD (2003b) Fos expression in spinothalamic and postsynaptic dorsal column neurons following noxious visceral and cutaneous stimuli. Pain 104:249–257

Panlilio LM, Christo PJ, Raja SN (2002) Current management of postherpetic neuralgia. Neurology 8:339–350

Pappagallo M, Haldey EJ (2003) Pharmacological management of postherpetic neuralgia. CNS Drugs 17:771–780

Pare M, Elde R, Mazurkiewicz JE, Smith AM, Rice FL (2001) The Meissner corpuscle revised: a multiafferented mechanoreceptor with nociceptor immunochemical properties. J Neurosci 21:7236–7246

Parent A (1996) Carpenter's human neuroanatomy, 9th edn. Williams and Wilkins, Baltimore

Patterson JT, Head PA, McNeill DL, Chung K, Coggeshall RE (1989) Ascending unmyelinated primary afferent fibers in the dorsal funiculus. J Comp Neurol 290:384–390

Patterson JT, Coggeshall RE, Lee WT, Chung K (1990) Long ascending unmyelinated primary afferent axons in the rat dorsal column: immunohistochemical localizations. Neurosci Lett 108:6–10

Paxinos G, Törk I, Halliday G, Mehler WR (1990) Human homologs to brainstem nuclei identified in other animals as revealed by acetylcholinesterase activity. In: Paxinos G (ed) The human nervous system. Academic Press, San Diego, pp 149–202

Pellegrini-Giampietro DE, Fan S, Ault B, Miller BE, Zukin RS (1994) Glutamate receptor gene expression in spinal cord of arthritic rats. J Neurosci 14:1576–1583

Penfield W, Boldrey E (1937) Somatic motor and sensory representation in the cerebral cortex of man as studied by electrical stimulation. Brain 60:389–443

Percheron G (2004) Thalamus. In: Paxinos G, Mai JK (2004) The human nervous system, 2nd edn. Elsevier, Amsterdam, pp 592–675

Perl ER (1996) Pain and the discovery of nociceptors. In: Belmonte C, Cervero F (eds) Neurobiology of nociceptors. Oxford University Press, Oxford, pp 5–36

Perry MJ, Lawson SN (1998) Differences in expression of oligosaccharides, neuropeptides, carbonic anhydrase and neurofilament in rat primary afferent neurons retrogradely labelled via skin, muscle or visceral nerves. Neuroscience 85:293–310

Peschanski M (1984) Trigeminal afferents to the diencephalon in the rat. Neuroscience 12:465–487

Peschanski M, Ralston HJ (1985) Light and electron microscopic evidence of transneuronal labelling with WGA-HRP to trace somatosensory pathways to the thalamus. J Comp Neurol 236:29–41

Peschanski M, Mantyh P, Besson JM (1983) Spinal afferents to the ventrobasal thalamic complex in the rat: an anatomical study using wheatgerm agglutinin conjugated to horseradish peroxidase. Brain Res 278:240–244

Petralia RS, Yokotani N, Wenthold RJ (1994) Light and electron micro-scope distribution of the NMDA receptor subunit NMDAR1 in the rat nervous system using a selective anti-peptide antibody. J Neurosci 14:667–696

Petralia RS, Wang YX, Mayat E, Wenthold RJ (1997) Glutamate receptor subunit 2-selective antibody shows a differential distribution of calcium-impermeable AMPA receptors among populations of neurons. J Comp Neurol 385:456–476

Petruska JC, Streit WJ, Johnson RD (1997) Localization of unmyelinated axons in rat skin and mucocutaneous tissue utilizing the isolectin GS-I-B4. Somatosens Mot Res 14:17–26

Petruska JC, Napaporn J, Johnson RD, Cooper BY (2002) Chemical responsiveness and histochemical phenotype of electrophysiologically classified cells of the adult rat dorsal root ganglion. Neuroscience 115:15–30

Peyron R, Laurent B, Garcia-Larrea L (2000) Functional imaging of brain responses to pain: a review and meta-analysis (2000). Neurophysiol Clin 30:263–288

Pezet S, Malcangio M, McMahon SB (2002) BDNF: a neuromodulator in nociceptive pathways? Brain Res Rev 40:240–249

Pfaller K, Arvidsson J (1988) Central distribution of trigeminal and upper cervical primary afferents in the rat studied by anterograde transport of horseradish peroxidase conjugated to wheat germ agglutinin. J Comp Neurol 268:91–108

Phend KD, Rustioni A, Weinberg RW (1995) An osmium-free method of Epon embedment that preserves both ultrastructure and antigenicity for post-embedding immunocytochemistry. J Histochem Cytochem 43:283–292

Pierret T, Lavallee P, Deshenes M (2000) Parallel streams for the relay of vibrissal information through thalamic barreloids. J Neurosci 20:7455–7462

Pin JP, Duvoisin R (1995) The metabotropic glutamate receptors: structure and functions. Neuropharmacology 34:1–26

Ploner M, Freund HJ, Schnitzler A (1999) Pain affect without pain sensation in a patient with postcentral lesion. Pain 81:211–214

Polgar E, Puskar Z, Watt C, Matesz C, Todd AJ (2002) Selective innervation of lamina I projection neurones that possess the neurokinin 1 receptor by serotonin-containing axons in the rat spinal cord. Neuroscience 109:799–899

Pomonis JD, Rogers SD, Peters CM, Ghilardi JR, Mantyh PW (2001) Expression and localization of endothelin receptors: implications for the involvement of peripheral glia in nociception. J Neurosci 21:999–1006

Popovich PG, Wei P, Stokes BT (1997) Cellular inflammatory response after spinal cord injury in Sprague-Dawley and Lewis rats. J Comp Neurol 377:443–464

Popratiloff A, Weinberg RJ, Rustioni A (1996a) AMPA receptor subunits underlying terminals of fine-caliber primary afferent fibers J Neurosci 16:3363–3372

Popratiloff A, Weinberg RJ, Rustioni A (1996b) AMPA receptor subunits in rat substantia gelatinosa after peripheral nerve injury. Soc Neurosci Abst 22:354

Popratiloff A, Rustioni A, Weinberg RJ (1997) Heterogeneity of AMPA receptors in the dorsal column nuclei of the rat. Brain Res 754:333–339

Popratiloff A, Weinberg RJ, Rustioni A (1998a) AMPA receptors at primary afferent synapses in substantia gelatinosa after sciatic nerve section. Eur J Neurosci 10:3220–3230

Popratiloff A, Weinberg RJ, Rustioni A (1998b) NMDAR1 and primary afferent terminals in the superficial spinal cord. Neuroreport 9:2423–2429

Porreca F, Ossipov MH, Gebhart GF (2002) Chronic pain and medullary descending facilitation. Trends Neurosci 25:319–325

Porro CA, Cetolo V, Francescato MP, Baraldi P (1998) Temporal and intensity coding of pain in human cortex. J Neurophysiol 80:3312–3320

Portenoy RK, Duma C, Foley KM (1986) Acute herpetic and postherpetic neuralgia: clinical review and current management. Ann Neurol 20:651–664

Price DD (2000) Psychological and neural mechanisms of the affective dimension of pain. Science 288:1769–1772

Price DD, Greenspan JD, Dubner R (2003) Neurons involved in the exteroceptive function of pain. Pain 106:215–219

Rainville P, Duncan GH, Price DD, Carrier B, Bushnell MC (1997) Pain affect encoded in human anterior cingulate but not somatosensory cortex. Science 277:968–971

Raja SN, Meyer RA, Ringkamp M, Campbell JN (1999) Peripheral neural mechanisms of nociception. In: Wall PD, Melzack R (eds) Textbook of pain. Churchill Livingstone, Edinburgh, pp 11–57

Ralston HJ (1979) The fine structure of laminae I, II and III of the macaque spinal cord. J Comp Neurol 184:643–684

Ralston HJ, Ralston DD (1979) The distribution of dorsal root axons in laminae I, II and III of the macaque spinal cord: a quantitative electron microscope study. J Comp Neurol 184:643–684

Ralston HJ, Ralston DD (1992) The primate dorsal spinothalamic tract: evidence for a specific termination in the posterior nuclei (Po/SG) of the thalamus. Pain 48:107–118

Ralston HJ, Ralston DD (1994) Medial lemniscal and spinal projections to the macaque thalamus: an electron microscopic study of differing GABAergic circuitry serving thalamic somatosensory mechanisms. J Neurosci 14:2485–2502

Ramer MS, Murphy PG, Richardson PM, Bisby MA (1998) Spinal nerve lesion-induced mechanoallodynia and adrenergic sprouting in sensory ganglia are attenuated in interleukin-6 knockout mice. Pain 78:115–121

Ramer MS, Thompson SW, McMahon SB (1999) Causes and consequences of sympathetic basket formation in dorsal root ganglia. Pain 6 (Suppl):S111–S120

Ramon y Cajal S (1909) Histologie du système nerveux de l'homme et des vertébrés. Tome premier: généralités, moelle, ganglion rachidiens, bulbe et protubérance. Maloine, Paris; (1972) Segunda reimpresion. Consejo Superior de Investigaciones Cientificas, Instituto Ramon y Cajal, Madrid

Ranson SW (1913) The course within the spinal cord of the non-myelinated fibers of the dorsal roots: a study of Lissauer's tract in the cat. J Comp Neurol 23:259–281

Rees H, Roberts MHT (1993) The anterior pretectal nucleus: a proposed role in the sensory processing. Pain 53:121–135

Regan JM, Peng P (2000) Neurophysiology of cancer pain. Cancer Control 7:111–119

Reshef E, Greenberg SB, Jankovic J (1985) Herpes zoster ophtalmicus followed by contralateral hemiparesis: report of two cases and review of literature. J Neurol Neurosurg Psychiat 48:122–127

Rexed B (1952) The cytoarchitectonic organization of the spinal cord in the cat. J Comp Neurol 96:415–495

Rexed B (1954) A cytoarchitectonic atlas of the spinal cord in the cat. J Comp Neurol 100:297–379

Rexed B (1964) Some aspects of the cytoarchitectonics and synaptology of the spinal cord. Prog Brain Res 2:58–92

Rexed B, Sourander P (1949) The caliber of central and peripheral neurites of the spinal ganglion cells and variations in fiber size at different levels of dorsal spinal roots. J Comp Neurol 91:297–306

Reynolds DV (1969) Surgery in the rat during electrical analgesia induced by focal brain stimulation. Science 164:444–445

Ribeiro-da-Silva A (1995) Substantia gelatinosa of spinal cord. In: Paxinos G (ed) The rat nervous system, 2nd edn. Academic Press, San Diego, pp 47–59

Ribeiro-da-Silva A, Coimbra A (1982) Two types of synaptic glomeruli and their distribution in laminae I–III of the rat spinal cord. J Comp Neurol 209:176–186

Ribeiro-da-Silva A, Coimbra A (1984) Capsaicin causes selective damage to type I synaptic glomeruli in rat substantia gelatinosa. Brain Res 290:380–383

Riddoch G (1938) The clinical features of central pain. Lancet 234:1093–1098, 1150–1156, 1205–1209

Rodriguez-Filho R, Santos ARS, Bertelli JA, Calixto JB (2003) Avulsion injury of the rat brachial plexus triggers hyperalgesia and allodynia in the hindpaws: a new model for the study of neuropathic pain. Brain Res 982:186–194

Rowbotham MC, Fields HL (1996) The relationship of pain, allodynia and thermal sensation in post-herpetic neuralgia. Brain 119:347–354

Rowbotham MC, Yosipovitch G, Connolly MK, Finlay D, Forde G, Fileds HL (1996) Cutaneous innervation density in the allodynic form of postherpetic neuralgia. Neurobiol Dis 3:205–214

Rozsa AJ, Beuerman RW (1982) Density and organization of free nerve endings in the corneal epithelium of the rabbit. Pain 14:105–120

Ruscheweyh R, Sandkühler J (2002) Role of kainite receptors in nociception. Brain Res Rev 40:215–222

Russo A, Conte B (1996) Afferent and efferent branching axons from the rat lumbo-sacral spinal cord project both to the urinary bladder and the urethra as demonstrated by double retrograde neuronal labelling. Neurosci Lett 219:155–158

Rustioni A (1973) Non-primary afferents to the nucleus gracilis from the lumbar cord of the cat. Brain Res 51:81–95

Rustioni A (1974) Non-primary afferents to the cuneate nucleus in the brachial dorsal funiculus of the cat. Brain Res 75:247–259

Rustioni A (1977) Spinal neurons project to the dorsal column nuclei of rhesus monkeys. Science 196:656–658

Rustioni A, Cuénod M (1982) Selective retrograde transport of D-aspartate in spinal interneurons and cortical neurons of rats. Brain Res 236:143–155

Rustioni A, Kaufman AB (1977) Identification of cells of origin of non-primary afferents to the dorsal column nuclei of the cat. Exp Brain Res 27:1–14

Rustioni A, Weinberg RJ (1989) The somatosensory system. In: Björklund A, Hökfelt T, Swanson LW (eds) Handbook of chemical neuroanatomy, vol 7. Elsevier, Amsterdam, pp 219–321

Rustioni A, Sanyal S, Kuypers HG (1971) A histochemical study of the distribution of the trigeminal divisions in the substantia gelatinosa of the rat. Brain Res 32:45–52

Rustioni A, Hayes NL, O'Neill S (1979) Dorsal column nuclei and ascending spinal afferents in macaques. Brain 102:95–125

Rybarova S, Kluchova D, Lovasova K, Kocisova M, Schmidtova K (2000) Expression of peptidergic and nitrergic structures in dorsal root ganglia of the rabbit. Eur J Histochem 44:377–384

Saab CY, Willis WD (2001) Nociceptive visceral stimulation modulates the activity of cerebellar Purkinje cells. Exp Brain Res 140:122–126

Saab CY, Willis WD (2003) The cerebellum: organization, functions and its role in nociception. Brain Res Rev 42:85–95

Saab CY, Kawasaki M, Al-Chaer ED, Willis WD (2001) Cerebellar cortical stimulation increases spinal visceral nociceptive responses. J Neurophysiol 85:2359–2363

Sabino MA, Mantyh PW (2005) Pathophysiology of bone cancer pain. J Support Oncol 3:15–24

Sabino MA, Luger NM, Mach DB, Rogers SD, Schwei MJ, Mantyh PW (2003) Different tumors in bone each give rise to a distinct pattern of skeletal destruction, bone cancer-related pain behaviors and neurochemical changes in the central nervous system. Int J Cancer 104:550–558

Safieh-Garabedian B, Poole S, Allchorne A, Winter J, Woolf CJ (1995) Contribution of interleukin-1 beta to the inflammation-induced increase in nerve growth factor levels and inflammatory hyperalgesia. Br J Pharmacol 115:1265–1275

Salt TE, Herrling PL (1995) Excitatory amino acid transmitter function in mammalian central pathways. In: Thomson AM, Wheal H (eds) Excitatory amino acids and synaptic transmission. Academic Press, London, pp 223–237

Sameda H, Takahashi Y, Takahashi K, Chiba T, Ohtori S, Moriya H (2003) Dorsal root ganglion neurones with dichotomising afferent fibres to both the lumbar disc and the groin skin. A possible neuronal mechanism underlying referred groin pain in lower lumbar disc diseases. J Bone Joint Surg Br 85:600–603

Sato J, Perl ER (1991) Adrenergic excitation of cutaneous pain receptors induced by peripheral nerve injury. Science 251:1608–1610

Sawamoto N, Honda M, Okada T, Hanakawa T, Kanda M, Fukiyama H, Konishi J, Shibasaki H (2000) Expectation of pain enhances responses to nonpainful somatosensory stimulation in the anterior cingulate cortex and parietal operculum/posterior insula: an event-related functional magnetic resonance imaging study. J Neurosci 20:7438–7445

Schäfer MKH, Nohr D, Krause JE, Weihe E (1993) Inflammation-induced upregulation of NK1 receptor mRNA in dorsal horn neurones. Neuroreport 4:1007–1010

Schaible HG, Schmidt RF (1983a) Activation of groups III and IV sensory units in medial articular nerve by local mechanical stimulation of knee joint. J Neurophysiol 49:35–44

Schaible HG, Schmidt RF (1983b) Responses of fine medial articular nerve to passive movements of knee joint. J Neurophysiol 49:1118–1126

Schaible HG, Schmidt RF (1985) Effects of experimental arthritis on the sensory properties of fine articular afferent units. J Neurophysiol 54:1109–1122

Schaible HG, Schmidt RF (1988) Time course of mechanosensitivity changes in articular afferents during a developing experimental arthritis. J Neurophysiol 60:2180–2195

Scharf JH (1958) Sensible Ganglien. In: Moellendorf v W, Bargmann W (eds) Handbuch der Mikroskopischen Anatomie des Menschen. Band 4 (Nervensystem), 3. Teil. Springer, Berlin, pp 1–408

Schmidbauer M, Budka H, Pilz P, Kurata T, Hondo R (1992) Presence, distribution and spread of productive varicella zoster virus infection in nervous tissues. Brain 115:383–398

Schmidt RF (1996) The articular polymodal nociceptor in health and disease. Prog Brain Res 113:53–81

Schmidt R, Schmelz M, Forster C, Ringkamp M, Torebjörk E, Handwerker H (1995) Novel classes of responsive and unresponsive C nociceptors in human skin. J Neurosci 15:333–341

Schmidt R, Schmelz M, Torebjörk HE, Handwerker HO (2000) Mechano-insensitive nociceptors encode pain evoked by tonic pressure to human skin. Neuroscience 98:793–800

Schnitzler A, Ploner M (2000) Neurophysiology and functional neuroanatomy of pain perception. J Clin Neurophysiol 17:592–603

Schoenen J, Faull RLM (1990) Spinal cord: cytoarchitectural, dendroarchitectural and myeloarchitectural organization. In: Paxinos G (ed) The human nervous system. Academic Press, San Diego, pp 19–54

Schoenen J, Faull RLM (2004) Spinal cord: cyto- and chemoarchitecture. In: Paxinos G, Mai JK (eds) The human nervous system, 2nd edn. Elsevier Academic Press, Amsterdam, pp 190–232

Schoenen J, Grant G (2004) Spinal cord: connections. In: Paxinos G, Mai JK (eds) The human nervous system, 2nd edn. Elsevier Academic Press, Amsterdam, pp 233–249

Schoepp DD, Conn PJ (1993) Metabotropic glutamate receptors in brain function and pathology. Trends Pharmacol Sci 14:13–20

Scholz J, Woolf CJ (2002) Can we conquer pain? Nat Neurosci 5 (Suppl):1062–1067

Schwartzman RJ, Maleki J (1999) Postinjury neuropathic pain syndromes. Med Clin North Am 83:597–626

Schwei MJ, Honore P, Rogers SD, Salak-Johnson JL, Finke MP, Ramnaraine ML, Clohisy DR, Mantyh PW (1999) Neurochemical and cellular reorganization of the spinal cord in a murine model of bone cancer pain. J Neurosci 19:10886–10897

Sebert ME, Shooter EM (1993) Expression of mRNA for neurotrophic factors and their receptors in the rat dorsal root ganglion and sciatic nerve following nerve injury. J Neurosci Res 36:357–367

Segond von Banchet G, Petrow PK, Bräuer R, Schaible HG (2000) Monoarticular antigen-induced arthritis leads to pronounced bilateral upregulation of the expression of neurokinin 1 and bradykinin 2 receptors in dorsal root ganglion neurons of rats. Arthritis Res 2:424–427

Seifert P, Spitznas M (2001) Tumours may be innervated. Virchows Arch 438:228–231

Seltzer Z, Dubner R, Shir Y (1990) A novel behavioral model of neuropathic pain disorders produced in rats by partial sciatic nerve injury. Pain 43:205–218

Sewards TV, Sewards M (2002) Separate, parallel sensory and hedonic pathways in the mammalian somatosensory system. Brain Res Bull 58:243–260

Shaw VE, Mitrofanis J (2001) Lamination of spinal cells projecting to the zona incerta of rats. J Neurocytol 30:695–704

Shi TJ, Holmberg K, Xu ZQ, Steinbusch H, de Vente J, Hökfelt T (1998) Effect of peripheral nerve injury on cGMP and nitric oxide synthase levels in rat dorsal root ganglia: time course and coexistence. Pain 78:171–180

Siddal PJ, Loeser JD (2001) Pain following spinal cord injury. Spinal Cord 39:63–73

Siddall PJ, Taylor DA, Cousins MJ (1997) Classification of pain following spinal cord injury. Spinal Cord 35:69–75

Siddal PJ, Taylor DA, McClelland JM, Rutkowski SB, Cousins MJ (1999) Pain report and the relationship of pain to physical factors in the first 6 months following spinal cord injury. Pain 81:187–197

Siddal PJ, Yezierski RP, Loeser JD (2000) Pain following spinal cord injury: clinical features, prevalence and taxonomy. IASP Newslett 3:3–7

Siegel P, Wepsic JG (1974) Alteration of nociception by stimulation of cerebellar structures in the monkey. Physiol Behav 13:189–194

Simmons Z, Feldman EL (2002) Update on diabetic neuropathy. Curr Opin Neurol 15:595–603

Slugg RM, Light AR (1994) Spinal cord and trigeminal projections to the pontine parabrachial region in the rat as demonstrated with Phaseolus vulgaris leucoagglutinin. J Comp Neurol 339:49–61

Smith FP (1978) Pathological studies of spinal nerve ganglia in relation to intractable intercostal pain. Surg Neurol 10:50–53

Snider WD, McMahon SB (1998) Tackling pain at the source: new ideas about nociceptors. Neuron 20:629–632

Snow PJ, Wilson P (1991) Plasticity in the somatosensory system of developing and mature mammals - the effects of injury to the central and peripheral nervous system. Progress in sensory physiology, vol 11. Springer, Berlin

Sommer EW, Kazimierczak J, Droz B (1985) Neuronal subpopulations in the dorsal root ganglion of the mouse as characterized by combination of ultrastructural and cytochemical features. Brain Res 346:310–326

Spiegel EA (1982) Relief of pain and spasticity by posterior column stimulation: a proposed mechanism. Arch Neurol 39:184–185

Spruce MC, Potter J, Coppini DV (2003) The pathogenesis and management of painful diabetic neuropathy: a review. Diabet Med 20:88–98

Stamford JA (1995) Descending control of pain. Br J Anaesth 75:217–227

Steel JH, Terenghi G, Chung JM, Na HS, Carlton SM, Polak JM (1994) Increased nitric oxide synthase immunoreactivity in rat dorsal root ganglia in a neuropathic pain model. Neurosci Lett 169:81–84

Steinbusch HWM (1981) Distribution of serotonin-immunoreactivity in the central nervous system of the rat—cell bodies and terminals. Neuroscience 6:557–618

Stern P, Edwards FA, Sakmann B (1992) Fast and slow components of unitary EPSCs on stellate cells elicited by focal stimulation in slices of rat visual cortex. J Physiol (Lond) 449:247–278

Stevens RT, Hodge CJ, Apkarian AV (1982) Kölliker-Fuse nucleus: the principal source of pontine catecholaminergic cells projecting to the lumbar spinal cord of cat. Brain Res 239:589–594

Stevens RT, Hodge CJ, Apkarian AV (1989) Medial, intralaminar, and lateral terminations of lumbar spinothalamic tract neurons: a fluorescent double-label study. Somatosens Mot Res 6:285–308

Stuesse SL, Cruce WL, Lovell JA, McBurney DL, Crisp T (2000) Microglial proliferation in the spinal cord of aged rats with a sciatic nerve injury. Neurosci Lett 287:121–124

Sugimoto T, Bennett GJ, Kajander KC (1989) Strychnine-enhanced transsynaptic degeneration of dorsal horn neurons in rats with an experimental painful peripheral neuropathy. Neurosci Lett 98:139–143

Sugimoto T, Bennett GJ, Kajander KC (1990) Transsynaptic degeneration in the superficial dorsal horn after sciatic nerve injury: effects of a chronic constriction injury, transection, and strychnine. Pain 42:205–213

Suzuki R, Morcuende S, Webber M, Hund SP, Dickenson AH (2002) Superficial NK1-expressing neurons control spinal excitability through activation of descending pathways. Nat Neurosci 5:1319–1326

Svensson P, Minoshima S, Beydoun A, Morrow TJ, Casey KL (1997) Cerebral processing of acute skin and muscle pain in humans. J Neurophysiol 78:450–460

Szekely JI, Torok K, Mate G (2002) The role of ionotropic glutamate receptors in nociception with special regard to the AMPA binding sites. Curr Pharm Des 8:887–912

Tachibana M, Wenthold RJ, Morioka H, Petralia RS (1994) Light and electron microscopic immunocytochemical localization of AMPA-selective glutamate receptors in the rat spinal cord. J Comp Neurol 344:413–454

Tajti J, Uddman R, Moller S, Sundler F, Edvinsson L (1999) Messenger molecules and receptor mRNA in the human trigeminal ganglion. J Auton Nerv Syst 28:176–183

Talbot JD, Marrett S, Evabs AC, Meyer E, Bushnell MC, Duncan GH (1991) Multiple representation of pain in human cerebral cortex. Science 251:1355–1358

Tamura E, Parry GJ (1994) Severe radicular pathology in rats with longstanding diabetes. J Neurol Sci 127:29–35

Tandrup T (1995) Are the neurons in the dorsal root ganglion pseudounipolar? A comparison of the number of neurons and number of myelinated and unmyelinated fibres in the dorsal root. J Comp Neurol 357:341–347

Tandrup T (2004) Unbiased estimates of number and size of rat dorsal root ganglion cells in studies of structure and cell survival. J Neurocytol 33:173–192

Tandrup T, Woolf CJ, Coggeshall RE (2000) Delayed loss of small dorsal root ganglion cells after transection of the rat sciatic nerve. J Comp Neurol 422:172–180

Tao F, Liaw WJ, Zhang B, Yaster M, Rothstein JD, Johns RA, Tao YX (2004) Evidence of neuronal excitatory amino acid carrier 1 expression in rat dorsal root ganglion neurons and their central terminals. Neuroscience 123:1045–1051

Tasker R (1990) Pain resulting from central nervous system pathology (central pain). In: Bonica JJ (ed) The management of pain. Lea and Fibiger, Philadelphia, pp 264–280

Terada T, Matsunaga Y (2001) S-100-positive nerve fibers in hepatocellular carcinoma and intrahepatic cholangiocarcinoma: an immunohistochemical study. Pathol Int 51:89–93

Terenghi G, Riveros-Moreno V, Hudson LD, Ibrahim NB, Polak JM (1993) Immunohistochemistry of nitric oxide synthase demonstrates immunoreactive neurons in spinal cord and dorsal root ganglia of man and rat. J Neurol Sci 118:34–37

Terenzi MG, Rees H, Morgan SJ, Foster GA, Roberts MHT (1991) The antinociception evoked by anterior pretectal nucleus stimulation is partially dependent upon ventrolateral medullary neurons. Pain 47:231–239

Terenzi MG, Rees H, Roberts MHT (1992) The pontine parabrachial region mediates some of the descending inhibitory effects of stimulating the anterior pretectal nucleus. Brain Res 594:205–214

Terenzi MG, Zagon A, Roberts MHT (1995) Efferent connections from the anterior pretectal nucleus to the diencephalon and mesencephalon in the rat. Brain Res 701:183–191

Thipeswamy T, Morris R (2001) Evidence that nitric oxide-induced synthesis of cGMP occurs in a paracrine but not an autocrine fashion and that the site of its release can be regulated: studies in dorsal root ganglia in vivo and in vitro. Nitric Oxide 5:105–115

Thipeswamy T, Morris R (2002) The roles of nitric oxide in dorsal root ganglion neurons. Ann N Y Acad Sci 962:103–110

Thomas PK, Lascelles G (1966) The pathology of diabetic neuropathy. Q J Med 35:489–509

Thompson SWN, Woolf CJ, Sivilotti LG (1993) Small caliber afferents produce a heterosynaptic facilitation of the synaptic responses evoked by primary afferent A fibers in the neonatal rat spinal cord in vitro. J Neurophysiol 69:2116–2128

Todd AJ (2002) Anatomy of primary afferents and projection neurones in the spinal dorsal horn with particular emphasis on substance P and the neurokinin 1 receptor. Exp Physiol 87:245–249

Todd AJ, Lewis SG (1986) The morphology of Golgi-stained neurons in lamina II of the rat spinal cord. J Anat 149:113–119

Todd AJ, Spike RC (1993) The localization of classical transmitters and neuropeptides within neurons in laminae I–III of the mammalian spinal dorsal horn. Prog Neurobiol 41:609–638

Todd AJ, Sullivan AC (1990) Light microscope study of the coexistence of GABA-like and glycine-like immunoreactivities in the spinal cord of the rat. J Comp Neurol 296:496–505

Todd AJ, Spike RC, Russel G, Johnston HM (1992) Immunohistochemical evidence that Met-enkephalin and GABA coexist in some neurons in rat dorsal horn. Brain Res 584:149–156

Todd AJ, Spike RC, Polgar E (1998) A quantitative study of neurons which express neurokinin 1 or somatostatin sst2a receptor in rat spinal dorsal horn. Neuroscience 85:459–473

Todd AJ, McGill MM, Shehab SA (2000) Neurokinin 1 receptor expression by neurons in laminae I, III and IV of the rat spinal dorsal horn that project to the brainstem. Eur J Neurosci 12:689–700

Todd AJ, Puskar Z, Spike RC, Hughes C, Watt C, Forrest L (2002) Projection neurons in lamina I of rat spinal cord with the neurokinin 1 receptor are selectively innervated by substance P-containing afferents and respond to noxious stimulation. J Neurosci 22:4103–4113

Tölle TR, Berthele A, Zieglgänsberger W, Seeburg PH, Wisden W (1993) The differential expression of 16 NMDA and non-NMDA receptor subunits in the rat spinal cord and in periaqueductal gray. J Neurosci 13:5009–5028

Tölle TR, Berthele A, Zieglgänsberger W, Seeburg PH, Wisden W (1995) Flip and flop variants of AMPA receptors in the rat lumbar spinal cord. Eur J Neurosci 7:1414–1419

Tominaga M, Caterina MJ, Malmberg AB, Rosen TA, Gilbert H, Skinner K, Raumann BE, Basmann AI, Julius D (1998) The cloned capsaicin receptor integrates multiple pain-producing stimuli. Neuron 21:1–20

Tong YG, Wang HF, Ju G, Grant G, Hökfelt T, Zhang X (1999) Increased uptake and transport of cholera toxin B-subunit in dorsal root ganglion neurons after peripheral axotomy: possible implications for sensory sprouting. J Comp Neurol 404:143–158

Torebjörk E (1997) Somatotopic organization along the central sulcus, for pain localization in humans, as revealed by positron emissions tomography. Exp Brain Res 117:192–199

Törk I, Hornung JP (1990) Raphe nuclei and the serotonergic system. In: Paxinos G (ed) The human nervous system. Academic Press, San Diego, pp 1001–1022

Tracey DJ (1995) Ascending and descending pathways in the spinal cord. In: Paxinos G (ed) The rat nervous system, 2nd edn. Academic Press, San Diego, pp 67–80

Tracey DJ, De Biasi S, Phend K, Rustioni A (1991) Aspartate-like immunoreactivity in primary afferent neurons. Neuroscience 40:673–686

Tran TD, Inui K, Hoshiyama M, Lam K, Kakigi R (2000) Conduction velocity of the spinothalamic tract following CO_2 laser stimulation of C-fibers in humans. Pain 95:125–131

Treede RD, Meyer RA, RajaSN, Campbell JN (1995) Evidence for two different heat transduction mechanisms in nociceptive primary afferents innervating monkey skin. J Physiol 483:747–758

Treede RD, Kenshalo DR, Gracely RH, Jones AKP (1999) The cortical representation of pain. Pain 79:105–111

Treede RD, Apkarian AV, Bromm B, Greenspan JD, Lenz FA (2000) Cortical representation of pain: functional characterization of nociceptive areas near the lateral sulcus. Pain 87:113–119

Truong H, McGinnis L, Dindo L, Honda CN, Giesler GJ (2004) Identification of dorsal root ganglion neurons that innervate the common bile duct of rats. Exp Brain Res 155:477–484

Tsuda M, Inoue K, Salter MW (2005) Neuropathic pain and spinal microglia: a big problem from molecules in small glia. Trends Neurosci 28:101–107

Tsuruoka M, Arai YC, Nomura H, Matsutani K, Willis WD (2003) Unilateral hindpaw inflammation induces bilateral activation of the locus coeruleus and the nucleus subcoeruleus in the rat. Brain Res Bull 61:117–123

Turnbull IM, Shulman R, Woodhurst WB (1980) Thalamic stimulation of neuropathic pain. J Neurosurg 52:486–493

Urban MO, Gebhart GF (1999) Supraspinal contributions to hyperalgesia. Proc Natl Acad Sci U S A 96:7687–7692

Usunoff KG, Marani E, Schoen JH (1997) The trigeminal system in man. Adv Anat Embryol Cell Biol 136:1–126

Usunoff KG, Kharazia VN, Valtschanoff JG, Schmidt HHHW, Weinberg RJ (1999) Nitric oxide synthase-containing projections to the ventrobasal thalamus in the rat. Anat Embryol 200:265–281

Valtschanoff JG, Weinberg RJ, Rustioni A, Schmidt HHHW (1992) Nitric oxide synthase and GABA colocalize in lamina II of rat spinal cord. Neurosci Lett 148:6–10

Valtschanoff JG, Phend KD, Bernardi PS, Weinberg RJ, Rustioni A (1994) Amino acid immunocytochemistry of primary afferent terminals in the rat dorsal horn. J Comp Neurol 346:237–252

Valtschanoff JG, Rustioni A, Guo A, Hwang SJ (2001) Vanilloid receptor VR1 is both presynaptic and postsynaptic in the superficial laminae of the rat dorsal horn. J Comp Neurol 436:225–235

Van Bockstaele EJ, Aston-Jones G, Pieriborne VA, Ennis M, Shipley M (1991) Subregions of the periaqueductal gray topographically innervate the rostral medulla in the rat. J Comp Neurol 309:305–327

Verge GM, Milligan ED, Maier SF, Watkins LR, Naeve GS, Foster AC (2004) Fractalkine (CX3CL1) and fractalkine receptor (CX3CR1) distribution in spinal cord and dorsal root ganglia under basal and neuropathic pain conditions. Eur J Neurosci 20:1150–1160

Verge VM, Xu Z, Xu XJ, Wiesenfeld-Hallin Z, Hökfelt T (1992) Marked increase in nitric oxide synthase mRNA in rat dorsal root ganglia after peripheral axotomy: in situ hybridization and functional studies. Proc Natl Acad Sci U S A 89:11617–11621

Vestergaard K, Nielsen J, Andersen G, Ingeman-Nielsen L, Jensen TS (1995) Sensory abnormalities in consecutive, unselected patients with central post-stroke pain. Pain 61:177–186

Vidnyanszky Z, Hamori J, Negyessy L, Ruegg D, Knopfel T, Kuhn R, Görcs T (1994) Cellular and subcellular localization of mGluR5a metabotropic glutamate receptor in rat spinal cord. Neuroreport 6:209–213

Villanueva L, Bouhassira D, Le Bars D (1996) The medullary subnucleus reticularis dorsalis (SRD) as a key link in both the transmission and modulation of pain signals. Pain 67:231–240

Villanueva L, Desbois C, Le Bars D, Bernard JF (1998) Organization of diencephalic projections from the medullary subnucleus reticularis dorsalis and the adjacent cuneate nucleus: a retrograde and anterograde tracer study in the rat. J Comp Neurol 390:133–160

Villar MJ, Wiesenfeld-Hallin Z, Xu XJ, Theodorsson E, Emson PC, Hökfelt T (1991) Further studies on galanin-, substance P-, and CGRP-like immunoreactivities in primary sensory neurons and spinal cord: effects of dorsal rhizotomies and sciatic nerve lesions. Exp Neurol 112:29–39

Vizzard MA, Erdman SL, Ericson VL, Stewart RJ, Roppolo JR, De Groat WC (1994) Localization of NADPH diaphorase in the lumbosacral spinal cord and dorsal root ganglia of the cat. J Comp Neurol 339:62–75

von Bokay J (1909) Über den ätiologischen Zusammenhang der Varizellen mit gewissen Fallen von Herpes Zoster. Wien Klin Wochenschr 22:1323–1326

Vrethem M, Boivie J, Arnqvist H, Holmgren H, Lindstrom T (2002) Painful polyneuropathy in patients with and without diabetes: clinical, neurophysiologic, and quantitative sensory characteristics. Clin J Pain 18:122–127

Waite PME, Tracey DJ (1995) Trigeminal sensory system. In: Paxinos G (ed) The rat nervous system, 2nd edn. Academic Press, San Diego, pp 705–724

Waite PME, Ashwell KWS (2004) Trigeminal sensory system. In: Paxinos G, Mai JK (eds) The human nervous system, 2nd edn. Elsevier Academic Press, Amsterdam, pp 1093–1124

Walker AE (1940) The spinothalamic tract in man. Arch Neurol Psychiat 43:284–298

Wall PD (1978) The gate control theory of pain mechanisms: a re-examination and re-statement. Brain 101:1–18

Wang CC, Westlund KN (2001) Responses of rat dorsal column neurons to pancreatic nociceptive stimulation. Neuroreport 12:2527–2530

Wang CC, Willis WD, Westlund KN (1999) Ascending projections from the area around the spinal cord central canal: A *Phaseolus vulgaris* leucoagglutinin study in rats. J Comp Neurol 415:341–367

Wang QP, Nakai Y (1994) The dorsal raphe: an important nucleus in pain modulation. Brain Res Bull 34:575–585

Wang XM, Yuan B, Hou ZL (1992) Role of the deep mesencephalic nucleus in the antinociception induced by stimulation of the anterior pretectal nucleus in rats. Brain Res 577:321–325

Watkins LR, Maier SF (2002a) The pain of being sick: implications of immune-to-brain communications for understanding pain. Annu Rev Psychol 51:29–57

Watkins LR, Maier SF (2002b) Beyond neurons: evidence that immune and glial cells contribute to pathological pain states. Physiol Rev 82:981–1011

Watkins LR, Milligan ED, Maier SF (2001) Glial activation: a driving force for pathological pain. Trends Neurosci 24:450–455

Watkins LR, Milligan ED, Maier SF (2003) Glial proinflammatory cytokines mediate exaggerated pain states: implications for clinical pain. Adv Exp Med Biol 521:1–21

Watson CP, Deck JH, Morshead C, Van der Kooy D, Evans RJ (1991) Post-herpetic neuralgia: further post-mortem studies of cases with and without pain. Pain 44:105–117

Weinberg RJ, Conti F, Van Eyck SL, Petrusz P, Rustioni A (1987) Glutamate immunoreactivity in the superficial laminae of rat dorsal horn and spinal trigeminal nucleus. In: Hicks TP, Lodge D, McLennan H (eds) Excitatory amino acid transmission. Neurology and Neurobiology, vol 24. Liss, New York, pp 126–133

Welch JM, Simon SA, Reinhart PH (2000) The activation mechanism of rat vanilloid receptor 1 by capsaicin involves the pore domain and differs from the activation by either acid or heat. Proc Natl Acad Sci U S A 97:13889–13894

West WL, Yeomans DC, Proudfit HK (1993) The function of noradrenergic neurons in mediating antinociception induced by electrical stimulation of the locus coeruleus in two different sources of Sprague-Dawley rats. Brain Res 626:127–135

Westlund KN, Coulter JD (1980) Descending projection of the locus coeruleus and subcoeruleus/medial parabrachial nuclei in monkey: axonal transport studies and dopamine-beta-hydroxylase immunochemistry. Brain Res Rev 2:235–264

Westlund KN, Craig AD (1996) Association of spinal lamina I projections with brainstem catecholamine neurons in the monkey. Exp Brain Res 110:151–162

Westlund KN, Bowker RM, Ziegler MG, Coulter JD (1983) Noradrenergic projections to the spinal cord of the rat. Brain Res 263:15–31

Westlund KN, Bowker RM, Ziegler MG, Coulter JD (1984) Origins and terminations of descending noradrenergic projections to the spinal cord of monkeys. Brain Res 292:1–16

Westlund KN, Carlton SM, Zhang D, Willis WD (1992) Glutamate-immunoreactive terminals synapse on primate spinothalamic tract cells. J Comp Neurol 322:519–527

Wiberg M, Westman J, Blomqvist A (1987) Somatosensory projection to the mesencephalon: an anatomical study in the monkey. J Comp Neurol 264:92–117

Wiech K, Preißl H, Birbaumer N (2001) Neuronale Netzwerke und Schmerzverarbeitung. Anaesthesist 55:2–12

Wiesenfeld-Hallin Z, Xu XJ (1998) Galanin in somatosensory functions. Ann N Y Acad Sci 863:383–389

Williams MN, Zahm DS, Jacquin MF (1994) Differential foci and synaptic organization of the principal and spinal trigeminal projections to the thalamus in the rat. Eur J Neurosci 6:429–453

Willis WD (1984) The raphe-spinal system. In: Barnes CD (ed) Brainstem control of spinal cord function. Research topics in physiology. Academic Press, New York, pp 141–214

Willis WD (1985) The pain system—the neural basis of nociceptive transmission in the mammalian nervous system. Karger, Basel

Willis WD (1992) Hyperalgesia and allodynia, Raven Press, New York

Willis WD (1997) Nociceptive functions of thalamic neurons. In: Steriade M, Jones EG, McCormick D (eds) Thalamus, vol. II. Experimental and clinical aspects. Elsevier, Amsterdam, pp 373–424

Willis WD (1999) Dorsal root potentials and dorsal root reflexes: a double-edged sword. Exp Brain Res 124:395–421

Willis WD (2001) Role of neurotransmitters in sensitization of pain responses. Ann N Y Acad Sci 933:142–156

Willis WD (2002) Long-term potentiation in spinothalamic neurons. Brain Res Rev 40:202–214

Willis WD, Coggeshall RE (1991) Sensory mechanisms of the spinal cord. Plenum Press, New York

Willis WD, Westlund KN (1997) Neuroanatomy of the pain system and of the pathways that modulate pain. J Clin Neurophysiol 14:2–31

Willis WD, Trevino DL, Coulter JD, Maunz RA (1974) Responses of primate spinothalamic tract neurons to natural stimulation of hindlimb. J Neurophysiol 37:358–372

Willis WD, Leonard RB, Kenshalo DR (1978) Spinothalamic tract neurons in the substantia gelatinosa. Science 202:986–988

Willis WD, Kenshalo DR, Leonard RB (1979) The cells of origin of the primate spinothalamic tract. J Comp Neurol 188:543–574

Willis WD, Westlund KN, Carlton SM (1995) Pain. In: Paxinos G (ed) The rat nervous system, 2nd edn. Academic Press, San Diego, pp 725–750

Willis WD, Zhang X, Honda CN, Giesler GJ (2001) Projections from the marginal zone and deep dorsal horn to the ventrobasal nuclei of the primate thalamus. Pain 92:267–276

Wilson P, Kitchener PD (1996) Plasticity of cutaneous primary afferent projections to the spinal dorsal horn. Prog Neurobiol 48:105–129

Woolard HH (1935) Observations on the terminations of cutaneous nerves. Brain 58:352–367

Woolf CJ, Doubell TP (1994) The pathophysiology of chronic pain—increased sensitivity to low threshold Aβ-fibre inputs. Curr Opin Neurobiol 4:525–534

Woolf CJ, Mannion RJ (1999) Neuropathic pain: aetiology, symptoms, mechanisms, and management. Lancet 353:1959–1964

Woolf CJ, Salter MW (2000) Neuronal plasticity: increasing the gain in pain. Science 288:1765–1768

Woolf CJ, Thompson SWN (1991) The induction and maintenance of central sensitization is dependent on N-methyl-D-aspartate acid receptor activation: implication for the treatment of post-injury pain hypersensitivity states. Pain 44:293–300

Woolf CJ, Shortland P, Coggeshall RE (1992) Peripheral nerve injury triggers central sprouting of myelinated afferents. Nature 355:75–78

Woolf CJ, Shortland P, Reynolds M, Ridings J, Doubell T, Coggeshall RE (1995) Reorganization of central terminals of myelinated primary afferents in the rat dorsal horn following peripheral axotomy. J Comp Neurol 360:121–134

Wree A, Itzev DE, Schmitt O, Usunoff KG (2005) Neurons in the dorsal column nuclei of the rat emit a moderate projection to the ipsilateral ventrobasal thalamus. Anat Embryol (in press)

Wu CL, Marsh A, Dworkin RH (2000) The role of sympathetic blocks in herpes zoster and postherpetic neuralgia. Pain 87:121–129

Wu J, Chen PX (1990) Cerebellar evoked potential elicited by stimulation of C-fiber in saphenous nerve of cat. Brain Res 522:144–146

Xu X, Fukuyama H, Yazawa SY, Mima T, Hanakawa T, Magata Y, Kanda M, Fujiwara N, Shindo K, Nagamine T, Shibasaki H (1997) Functional localization of pain perception in the human brain studied by PET. Neuroreport 8:555–559

Yamashiro T, Satoh K, Nakagawa K, Moriyama H, Yagi T, Takada K (1998) Expression of Fos in the rat forebrain following experimental tooth movement. J Dent Res 77:1920–1925

Yang Y, Ozawa H, Lu H, Yuri K, Hayashi S, Nihonyanagi K, Kawata M (1998) Immunocytochemical analysis of sex differences in calcitonin gene-related peptide in the rat dorsal root ganglion, with special reference to estrogen and its receptor. Brain Res 791:35–42

Yaszay B, Jablecki CK, Safran MR (2000) Zoster paresis of the shoulder. Case report and review of the literature. Clin Orthop 377:112–118

Yeomans DC, Proudfit HK (1992) Antinociception induced by microinjection of substance P into the A7 catecholamine cell group in the rat. Neuroscience 49:681–691

Yezierski RP (1988) Spinomesencephalic tract: projections from the lumbosacral spinal cord of the rat, cat, and monkey. J Comp Neurol 267:131–146

Yezierski RP, Mendez CM (1991) Spinal distribution and collateral projections of rat spinomesencephalic tract cells. Neuroscience 44:113–130

Yezierski RP, Gerhart KD, Schrock BJ, Willis WD (1983) A further examination of effects of cortical stimulation in primate spinothalamic tract cells. J Neurophysiol 49:424–441

Yezierski RP, Kaneko T, Miller KE (1993) Glutaminase-like immunoreactivity in rat spinomesencephalic tract cells. Brain Res 624:304–308

Yoshida A, Sessle BJ, Dostrovsky JO, Chiang CY (1992) Trigeminal and dorsal column nuclei projections to the anterior pretectal nucleus in the rat. Brain Res 590:81–94

Yoshimura M, Jessell T (1990) Amino acid-mediated EPSPs at primary afferent synapses with substantia gelatinosa neurones in the rat spinal cord. J Physiol 430:315–335

Yu XH, Zhang ET, Craig AD, Shigemoto R, Ribeiro-da-Silva A, De Koninck Y (1999) NK-1 receptor immunoreactivity in distinct morphological types of lamina I neurons of the primate spinal cord. J Neurosci 19:3545–3555

Zaal MJW, Völker-Dieben HJ, D'Amaro J (2000) Risc and prognostic factors of postherpetic neuralgia and focal sensory denervation: a prospective evaluation in acute herpes zoster ophthalmicus. Clin J Pain 16:345–351

Zagon A, Terenzi MG, Roberts MHT (1995) Direct projections from the anterior pretectal nucleus to the ventral medulla oblongata in rats. Neuroscience 65:253–272

Zenker W, Neuhuber WL (eds) (1990) The primary afferent neuron—a survey of recent morpho-functional aspects. Plenum Press, New York

Zhang C, Yang SW, Guo YG, Qiao JT, Dafny N (1997) Locus coeruleus stimulation modulates the nociceptive response in parafascicular neurons: an analysis of descending and ascending pathways. Brain Res Bull 42:273–278

Zhang D, Carlton SM, Sorkin LS, Willis WD (1990) Collaterals of primate spinothalamic tract neurons to the periaqueductal gray. J Comp Neurol 296:277–290

Zhang D, Owens CM, Willis WD (1991) Two forms of inhibition of spinothalamic tract neurons produced by stimulation of the periaqueductal gray and cerebral cortex. J Neurophysiol 65:1567–1579

Zhang ET, Craig AD (1997) Morphology and distribution of spinothalamic lamina I neurons in the monkey. J Neurosci 17:3274–3284

Zhang ET, Han ZS, Craig AD (1996) Morphological classes of spinothalamic lamina I neurons in the cat. J Comp Neurol 367:537–549

Zhang JD, Yang XL (1999) Projections from subnucleus oralis of the spinal trigeminal nucleus to contralateral thalamus via the relay of the juxtatrigeminal nucleus and dorsomedial part of the principal sensory trigeminal nucleus in the rat. J Hirnforsch 39:301–310

Zhang X, Ju G, Elde R, Hökfelt T (1993a) Effect of peripheral nerve cut on neuropeptides in dorsal root ganglia and the spinal cord of monkey with special reference to galanin. J Neurocytol 22:342–381

Zhang X, Verge V, Wiesenfeld-Hallin Z, Ju G, Bredt D, Snyder SH, Hökfelt T (1993b) Nitric oxide synthase-like immunoreactivity in lumbar dorsal root ganglia and spinal cord of rat and monkey and effect of peripheral axotomy. J Comp Neurol 335:563–575

Zhang X, Bean AJ, Wiesenfeld-Hallin Z, Xu XJ, Hökfelt T (1995a) Ultrastructural studies on peptides in the dorsal horn of the rat spinal cord. III. Effects of peripheral axotomy with special reference to galanin. Neuroscience 64:893–915

Zhang X, Bean AJ, Wiesenfeld-Hallin Z, Hökfelt T (1995b) Ultrastructural studies on peptides in the dorsal horn of the rat spinal cord. IV. Effects of peripheral axotomy with special reference to neuropeptide Y and vasoactive intestinal polypeptide/peptide histidine isoleucine. Neuroscience 64:917–941

Zhang X, Kostarczyk E, Giesler GJ (1995c) Spinohypothalamic tract neurons in the cervical enlargement of rats: descending axons in the ipsilateral brain. J Neurosci 15:8393–8407
Zhang X, Bao L, Shi TJ, Ju G, Elde R, Hökfelt T (1998) Down-regulation of mu-opioid receptors in rat and monkey dorsal root ganglion neurons and spinal cord after peripheral axotomy. Neuroscience 82:223–240
Zhang X, Wenk HN, Gokin AP, Honda CN, Giesler GJ (1999) Physiological studies of spinohypothalamic tract neurons in the lumbar enlargement of monkeys. J Neurophysiol 82:1054–1058
Zhou XF, Deng YS, Chie E, Xue Q, Zhong JH, McLachlan EM, Rush RA, Xian CJ (1999) Satellite-cell-derived nerve growth factor and neurotrophin-3 are involved in noradrenergic sprouting in the dorsal root ganglia following peripheral nerve injury in the rat. Eur J Neurosci 11:1711–1722

Subject Index

MIX
Papier aus verantwortungsvollen Quellen
Paper from responsible sources
FSC® C105338

If you have any concerns about our products,
you can contact us on
ProductSafety@springernature.com

In case Publisher is established outside the EU,
the EU authorized representative is:
Springer Nature Customer Service Center GmbH
Europaplatz 3, 69115 Heidelberg, Germany

Printed by Libri Plureos GmbH
in Hamburg, Germany